JEANETTE WINTERSON

Jeanette Winterson CBE was born in Manchester.
Adopted by Pentecostal parents she was raised to
be a missionary. This did and didn't work out.

Discovering early the power of books she left
home at sixteen to live in a Mini and get on with
her education. After graduating from Oxford
University she worked for a while in the theatre
and published her first novel, *Oranges Are Not the
Only Fruit*, at twenty-five. She has written eleven
novels for adults, as well as children's books, non-
fiction and screenplays. She is Professor of New
Writing at the University of Manchester. She lives
in the Cotswolds in a wood and in Spitalfields,
London.

She believes that art is for everyone and it is her
mission to prove it.

ALSO BY JEANETTE WINTERSON

Novels

Oranges Are Not the Only Fruit
The Passion
Sexing the Cherry
Written on the Body
Art & Lies
Gut Symmetries
The Powerbook
Lighthousekeeping
The Stone Gods
The Gap of Time
Frankissstein

Comic Books

Boating for Beginners

Short Stories

The World and Other Places
Midsummer Nights (ed.)
Christmas Days: 12 Stories and
12 Feasts for 12 Days

Novellas

Weight (Myth)
The Daylight Gate (Horror)

Non-fiction

Art Objects: Essays in Ecstasy
and Effrontery
Why Be Happy When You Could
Be Normal?
Courage Calls to Courage
Everywhere

Collaborations

LAND (with Antony Gormley
and Clare Richardson)

Children's Books

Tanglewreck
The Lion, the Unicorn
and Me
The King of Capri
The ..... of the Sun

JEANETTE WINTERSON

12 Bytes

How Artificial Intelligence Will Change
the Way We Live and Love

VINTAGE

1 3 5 7 9 10 8 6 4 2

Vintage is part of the Penguin Random House group of companies
whose addresses can be found at global.penguinrandomhouse.com

Penguin
Random House
UK

Copyright © Jeanette Winterson 2021
Postscript copyright © Jeanette Winterson 2022

Jeanette Winterson has asserted her right to be identified as the author of this
Work in accordance with the Copyright, Designs and Patents Act 1988

First published in Vintage in 2022
First published in hardback by Jonathan Cape in 2021

penguin.co.uk/vintage

A CIP catalogue record for this book is available from the British Library

ISBN 9781529112979

Printed and bound in Great Britain by Clays Ltd, Elcograf S.p.A.

The authorised representative in the EEA is Penguin Random House Ireland,
Morrison Chambers, 32 Nassau Street, Dublin D02 YH68

Penguin Random House is committed to a sustainable future
for our business, our readers and our planet. This book is made
from Forest Stewardship Council® certified paper.

These essays are for my godchildren, Ellie and Cal Shearer. Lucy Reynolds. Lucy is studying the past in order to understand the present and the future. (That is what the humanities are for.) Ellie is writing her own books now. Cal is in a lab in Oxford, building a brain.

Contents

ZONE FOUR: The Future
How the Future will be Different to the Past – and How It Won't.

How These Essays Came About

In 2009 – 4 years after it was published – I read Ray Kurzweil's *The Singularity Is Near*. It is an optimistic view of the future – a future that depends on computational technology. A future of superintelligent machines. It is also a future where humans will transcend our present biological limits.

I had to read the book twice – once for the sense and once for the detail.

After that, just for my own interest, year-in, year-out, I started to track this future; that meant a weekly read through *New Scientist*, *Wired*, the excellent technology pieces in the *New York Times* and the *Atlantic*, as well as following the money via the *Economist* and *Financial Times*. I picked up any new science and tech books that came out, but it wasn't enough for me. I felt I wasn't seeing the bigger picture.

How did we get here?

Where might we go?

I am a storyteller by trade – and I know everything we do is a fiction until it's a fact: the dream of flying, the dream of space travel, the dream of speaking to someone instantly, across time and space, the dream of not dying – or of returning. The dream of life-forms, not human, but alongside the human. Other realms. Other worlds.

*

Long before I read Ray Kurzweil, I read Harold Bloom, the American Jewish literary critic, whose pursuit of excellence was relentless. One of his more private books – in that he was unravelling something for himself – is *The Book of J* (1990), where Bloom looks at the earliest texts that were later redacted and varnished to become the Hebrew Bible. The first 5 books, the Pentateuch, were written around 10 centuries before the birth of the man Jesus – so they are separated from us by around 3,000 years.

Bloom thinks that the author of those early texts was a woman, and Bloom was certainly no feminist. His arguments are persuasive and it delights me that the most famous character in Western literature – God, the Author of All – was himself authored by a woman.

In the exploration of this story, Bloom offers his own translation of the Blessing – the Blessing promised by Yahweh to Israel – but really, the blessing any of us would want. And it isn't 'Be fruitful and multiply' – that's a command, not a blessing. It is this: More Life into a Time Without Boundaries.

Isn't that what computing technology will offer?

*

Bloom points out that most humans are fixated on space without boundaries. Think about it: land-grab, colonisation, urban creep, loss of habitat, the current fad for seasteading (sea cities with vast oceans at their disposal).

And space itself – the go-to fascination of rich men: Richard Branson, Elon Musk, Jeff Bezos.

When I think about artificial intelligence, and what is surely to follow – artificial general intelligence, or superintelligence – it seems to me that what this affects most, now and later, isn't space but time.

The brain uses chemicals to transmit information. A computer uses electricity. Signals travel at high speeds through the nervous system (neurons fire 200 times a second, or 200 hertz) but computer processors are measured in gigahertz – billions of cycles per second.

We know how fast computers are at calculation – that's how it all started, back in Bletchley Park in World War Two, when the human teams just couldn't calculate fast enough to crack the German Enigma codes. Computers use brute force to process numbers and data. In time terms, they can get through more, faster.

Acceleration has been the keyword in our world since the Industrial Revolution. Machines use time differently to humans. Computers are not time-bound. As biological beings, humans are subject to time, most importantly our allotted span: we die.

And we hate it.

One of the near-future breakthroughs humans can expect is to live longer, healthier lives, perhaps much longer, even 1,000-year lives, if AI biologist Aubrey de Grey is right. Rejuvenation biotechnology will aim to slow down the accumulation of ageing damage in our organs and tissues, as well as repairing or replacing what is no longer fit for purpose.

More life into a time without boundaries.

And if that doesn't work there is always the possibility of brain upload, where the contents of your brain are transferred to another platform – initially not made of meat.

Would you choose that?

What if dying is a choice?

Living long, perhaps living forever, will certainly affect our notions of time – but recall that clock-time is really only an invention/necessity of the Machine Age. Animals don't live on clock-time, they live seasonally. Humans will find new ways of measuring time.

I wanted to think about the start of the Machine Age – the Industrial Revolution, and its impact on humans. I come from Lancashire, where those first, vast, cotton-processing factories changed life on earth for everyone. It is so near in time – only 250 years – how did we get where we are now?

I wanted to know why so few women seem to be interested in computing science. Was it always the case?

And I wanted to get a bigger picture of AI, by considering religion, philosophy, literature, myth, art, the stories we tell about human life on earth, our sci-fi, our movies, our enduring fascination/intuition that there might be more going on – whether it's ET, aliens or angels.

Artificial intelligence was the term coined in the mid-1950s by John McCarthy – an American computing expert – who, like his friend Marvin Minsky, believed computers could achieve human levels of intelligence by the 1970s. Alan Turing had thought the year 2000 was realistic.

Yet from the coining of a term – AI – 40 years would pass before IBM's Deep Blue beat Kasparov at chess in 1997. That's because computational power is the sum of computer storage (memory) and processing speed. Simply, computers weren't powerful enough to do what McCarthy, Minsky and Turing knew they would be able to do. And before those men, there was Ada Lovelace, the early-19th century genius who inspired Alan Turing to devise the Turing Test – when we can no longer tell the difference between AI and bio-human.

We aren't there yet.

Time is hard to gauge.

These 12 bytes are not a history of AI. They are not the story of Big Tech or Big Data, though we often meet on that ground.

A bit is the smallest unit of data on a computer – it's a binary digit, and it can have a value of 0 or 1. 8 bits make a byte.

My aim is modest; I want readers who imagine they are not much interested in AI, or biotech, or Big Tech, or data-tech, to find that the stories are engaging, sometimes frightening, always connected. We all need to know what's going on as humans advance, perhaps towards a transhuman – or even a post-human – future.

There are some repetitions in these essays; they are pieces of a jigsaw, but they also stand alone.

Of course, if our relationship to time changes, then our relationship to space changes too – because Einstein demonstrated that time and space are not separate, but part of the same fabric.

Humans love separations – we like to separate ourselves from other humans, usually in hierarchies, and we separate ourselves from the rest of biology by believing in our superiority. The upshot is that the planet is in peril, and humans will fight humans for every last resource.

Connectivity is what the computing revolution has offered us – and if we could get it right, we could end the delusion of separate silos of value and existence. We might end our anxiety about intelligence. Human or machine, we need all the intelligence we can get to wrestle the future out of its pact with death – whether war, or climate breakdown, or probably both.

Let's not call it artificial intelligence. Perhaps alternative intelligence is more accurate. And we need alternatives.

Zone One

The Past

How We Got Here. A Few Lessons From History.

Love(Lace) Actually

At the beginning of the future were two young women: Mary Shelley and Ada Lovelace.

Mary was born in 1797. Ada was born in 1815.

Each of these young women tore their way into history in the early years of the Industrial Revolution. The start of the Machine Age.

Both women belonged to their own time – as we all do – and both women were flares flung across time, throwing light on the world of the future. The world that is our present day. A world that is set on course to change the nature, role, and, perhaps, the dominance of Homo sapiens. History repeats itself – the same struggles in different disguises – but AI is new to human history. In their different ways the young women saw it coming.

Mary Shelley wrote the novel *Frankenstein* when she was 18. In that story, the doctor-scientist Victor Frankenstein builds an oversize, humanoid creature, using body parts and electricity.

Electricity, as a force that could be harnessed for our purposes, was poorly understood, and not in use in any practical way.

Read *Frankenstein* now, and it's more than an early example of a novel by a woman, more than a Gothic novel, or a novel about motherless children or the importance of education for everyone. It's more than sci-fi, more than the world's most famous monster; it's a message in a bottle.

Open it.

We are the first generation since that book was published, over 200 years ago, that is also beginning to create new life-forms. Like Victor Frankenstein's, our digital creations depend on electricity – but not on the rotting discards of the graveyard. Our new intelligence – embodied or non-embodied – is built out of the zeros and ones of code.

And that's where we meet Ada, the world's first computer programmer – of a computer that hadn't been built.

Both Mary and Ada intuited that the upheavals of the Industrial Revolution would lead to more than the development and application of machine technology. They recognised a decisive shift in the fundamental framing of what it means to be human.

Victor Frankenstein: 'If I could bestow animation upon lifeless matter …'

Ada: 'An explicit function … worked out by the engine … without having been worked out by human heads and human hands first.'

Mary and Ada never met but they had someone crucial in common.

Lord Byron was at that time England's most famous living poet. He was dashing, rich, and young. Hounded by scandal and divorce in England, in 1816 he proposed a holiday to Lake Geneva, with his great friend, the poet Percy Bysshe Shelley, Shelley's wife, Mary, and Mary's stepsister, Claire Clairmont, by now Byron's mistress.

The holiday was a success, until it started to rain torrentially, and the young people could not go out. Byron suggested that they alleviate the monotonous indoor days by each writing a story of the supernatural. Mary Shelley began the dark, rain-soaked prophecy that became *Frankenstein*.

Byron himself couldn't think of a story. He was irritable and distracted, in part by the legal battles over his divorce, and the settlement for his new child.

Byron wrote flurries of letters about his daughter's upbringing, but he had left England, never to return, and he did not see his daughter again.

Her name was Ada.

Ada's mother, Annabella Wentworth, was a devout Christian; one of many reasons why her marriage to bisexual Byron was never going to work.

Annabella had money, and status, but at that time women and children were legally the property of their nearest male relative. Even after there was a Deed of Separation between the two of them, Byron's wishes for his child had legal weight. His long, written instructions about his daughter's education included, above all, that she was not to be led astray by poetry.

This suited Ada's mother rather well. The last thing she wanted in her life was more of the Byronic temperament. A talented amateur mathematician herself, she engaged maths tutors for tiny Ada, in order to correct any inherited poetical leanings and to dilute the effects of Byronic blood. Not for nothing was Byron called 'mad, bad and dangerous to know'.

As it happened, little Ada was delighted by numbers. This was at a time when even the wealthiest women were not educated beyond reading, writing, drawing, playing the piano and possibly learning French or German. Women did not go to school.

Mary Shelley's mother – Mary Wollstonecraft – had written passionately about the importance of education for women in her radical *A Vindication of the Rights of Woman* (1792), and it is not an accident that Victor Frankenstein fails to educate his monster, leaving him to be self-taught. Women at that time had to teach themselves Latin and Greek, mathematics and the natural sciences, all the 'masculine' subjects their brothers could expect to be taught at school. The assumption was that women didn't have the brains for serious study – and when they did have the brains, too much concentration made them crazy, ill, or lesbian.

On their holiday at Lake Geneva, Mary Shelley spent plenty of time arguing with Byron about gender. Byron was disappointed that the 'glorious boy' he had longed for had turned out to be a glorious girl. He didn't live long enough to see her become a maths-head.

One of Ada's maths tutors, Augustus De Morgan, was worried that too much maths might break Ada's delicate constitution. At the same time, he believed her to be more gifted and able than any pupil (read *boys*) he had taught, and said in a letter to her mother that Ada could become 'an original mathematical investigator, perhaps of first-rate eminence'.

Poor Ada. Told to study maths to avoid going poetically mad. Then told she was at risk of going mathematically mad.

None of this mattered to Ada, who seems to have known her own mind from her early years.

At 17 she was invited to a party at 1 Dorset Street, London. The home of Charles Babbage.

Babbage was independently wealthy, clever, eccentric, and had persuaded the British Government to grant him £17,000 (about £1.7 million now) to build what he called his Difference Engine: a crank-handled adding machine designed to calculate and print the logarithmic tables used by engineers, sailors, accountants, machine builders – anybody who wanted to do the calculations speedily by using pre-printed tables.

His idea, like so many innovations of the Industrial Revolution, was to mechanise repetitive work. The word 'computer' at this time was used for human operators doing the tedious arithmetical tabling that Babbage imagined (correctly) could be done by his Difference Engine.

Babbage was Lucasian Professor of Mathematics at Cambridge, a post held by both Isaac Newton before him and Stephen Hawking after him (and still never held by a woman, btw). Babbage had a fascination for mechanical automata, as well as numbers. Building a cogs-and-wheels calculating machine was perfect for him.

And for Ada, as it turned out.

To be invited to a Babbage party you had to be beautiful, clever or aristocratic. Sacks of cash couldn't get you through the door. Ada wasn't a society beauty (thank God), but she was clever and her father (whether he liked it or not) was Lord Byron.

At 17, Ada was in.

A working section of Babbage's Difference Engine was on display in his drawing room. Ada was fascinated by it, and, as the party hummed and buzzed around them, Ada and Babbage

played with the machine. Babbage was so excited that he lent her the plans.

Suddenly, the tricky and difficult 40-something genius who couldn't do small talk, and hated barrel organs, had found a friend who understood his work, both practically and conceptually.

The two of them began to exchange letters while Ada carried on with her mathematical studies. Whether or not this meeting with Ada inspired Babbage to go further, he began that year to put together a new kind of calculating device, which he called the Analytical Engine, and this device was the world's first non-human computer.

Even though it was never built.

Babbage realised that the punched-cards system used on the mechanical Jacquard loom could be used to self-operate a calculating machine. No need for a crank handle. The calculating machine could also use the punched cards to store memory. This was an extraordinary insight.

*

Punched cards are stiff cards with holes in them. The Frenchman Joseph-Marie Jacquard patented a mechanism in 1804 that allowed the pattern of a piece of cloth to be expressed as a series of holes on a card. This was a genius moment of abstract intuition – closer to the quantum-mechanical patterned universe than the 3D realism of the Industrial Revolution. It makes sense that Babbage grasped its implications for computing. Actually, it makes no sense – it was a mental leap for both men.

On a Jacquard loom, the arrangement of the holes deter-mines the pattern. Using this system meant there was no need for a master weaver to pass the weft thread laboriously under the warp thread to weave the cloth and make the pattern. It is the order of warp and weft that sets the pattern. This is skilled but repetitive work, and, as with so many of the innovations of the Industrial Revolution, by mechanising the repetition, there was no longer any need for the same level of human skill. Mechanising repetition is an engineering challenge, but engi-neering alone isn't the key leap of the Jacquard loom: the leap is seeing what is solid and tangible as a series of holes (in effect, empty space) arranged as a pattern.

Punched cards were used in the earliest commercial tabulators, and later on in early computers. They remained in use (holes in tape), as feed-in computer programmes, until the mid-1980s. Babbage didn't patent this idea – he was a terrible businessman. The punched-card system was patented in 1894 by an American entrepreneur called Herman Hollerith. Hollerith was the son of a German immigrant. His Tabulating Machine Company even-tually became IBM in 1924. IBM stands for International Business Machine.

(You can see why the names Difference Engine and Analyt-ical Engine were never going to cut it in the marketplace.)

*

Ada was thrilled with the punched-card idea. She wrote: 'The Analytical Engine weaves algebraic patterns just as the Jacquard loom weaves flowers and leaves.'

Except that it didn't, because Babbage could never quite finish his engine, or even nearly finish it – and so the cogs, levers, pistons, arms, screws, wheels, rack and pinions gears, bevels, studs, springs, and punch cards lived in a kind of Steampunk Victoriana, where everything was massive, solid, dimensional (think railways, iron ships, factories, piping, track, cylinders, furnaces, metal, coal), but at the same time a thought-experiment fantasy. For Babbage and Ada, imagining what could happen meant that it *had* happened – and in the most important way they were right. The future had been imagined but the weight of the present was too heavy for it. Fun though it was to play at building a coal-fired, steam-powered, punched-card computer made out of tons of metal, this was not the answer to the instant and elegant universe of numbers where Ada and Babbage lived.

But elegance was still a long way off.

In 1944 (not 1844) the world's first electronic digital computer – Colossus – built by a British team in World War Two, and housed in Bletchley Park, measured 7ft high by 17ft wide and 11ft deep. It weighed 5 tonnes, was made out of 2,500 valves, 100 Logic Gates and 10,000 resistors connected by 7km of wiring. Actually, the existence of this computer set was kept secret until the 1970s, so the American-built ENIAC (Electronic Numerical Integrator And Computer) often takes the title of 'first', arriving in 1946.

Babbage would have loved the Colossus set. And the punched-paper tapes. In fact, if Babbage and Ada had jumped in a time machine and got out in the year 1944 they would have been astonished by motor vehicles, rubber boots, radios, telephones, aeroplanes, even zips, but one look at the Colossus set and they would have *recognised* it.

There is still a fair bit of mansplaining around Ada: that she was just a hanger-on, that her maths was wonky, that she didn't really write the set of notes that explained the Analytical Engine. That she over-estimated herself, that she was vain, and that Babbage indulged her.

This is the Brontë Zone. Remember the theory that the drunken runt of a brother, Branwell, wrote everything, or at least *Wuthering Heights*? That theory is delightfully lampooned through the character of Mr Mybug, in Stella Gibbons' novel *Cold Comfort Farm*.

Strangely, not strangely(?), Brontë Zone and Ada Zone mansplaining is still alive and well on the web.

More accurately, and more importantly, there is now Ada Lovelace Day, celebrated on the second Tuesday of October. In the UK we have the Ada Lovelace Institute (founded 2018), an independent body whose mission is to ensure that data use and AI technology work for the benefit of society as a whole – and not for the self-entitled few.

As a female mathematician, Ada stands as a beacon for women in maths and computing. Women need a beacon, because external and internal prejudices are still running full throttle. Right now, in the first quarter of the 21st century, only about 20% of the people working in electrical engineering, computer programming and machine learning are female.

Women don't build the platforms or write the programmes. Even fewer women are in tech start-ups. Why is all this vital, future-facing technology male-dominated?

Explanation? Well, take your pick.

There's the Mars vs Venus version that it's a gender thing: women don't really get computing science. (Different brain? Hormones?)

There's the Equality version: women don't *want* to do this kind of thing. There's nothing in a woman's way, of course, not now. Women are *so* welcome ... Free choice, girls!

There's the Slow Progress version: until girls are encouraged to take maths, computing science and tech more seriously at school (instead of social media encouraging videos on How To Do Your Make-Up Like Kim Kardashian) they can't be parachuted into challenging tech jobs just to tick-box gender parity.

All these versions overlook the fact (FACT!) that many thousands of women worked as human computers (doing the calculations by hand), and as computer programmers, from World War Two onwards. The fabulous story of Katherine Johnson and the African-American women she worked with at NASA became the movie *Hidden Figures* (2016).

The six women who programmed one of the world's first viable programmable computers – the ENIAC – were not invited to its launch at the University of Pennsylvania in 1946, nor were they mentioned in the celebrations. Here's a photo of Betty Jennings and Frances Bilas operating the ENIAC:

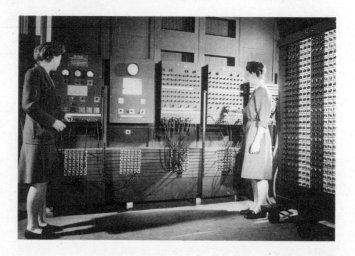

*

These six women were drawn from a pool of 200 women working as 'human' computers. This work was categorised as 'clerical'. It wasn't. But it was a way of keeping women in their (low-paid) place.

Women are as smart as men. I am writing this self-evident proposition because the way the world is, it is not self-evident.

Many women find maths easy – or at least do-able. Yet an engineering professor I know at the University of Manchester, UK, told me that a boy with a grade B at Maths A Level will often go on to do engineering. A girl with an A probably won't.

So, what's going on?

I don't think that the answer to the question, *Why aren't lots more women in computing science?* has much to do with female brains or female hormones, or even female free choice.

Gender is still the roadblock in the way.

That includes the way we manage gender in the workplace. Is an all-male tech-room a comfortable place for a woman to be? Do women routinely see other women working in tech? We copy what we can see – and a truly mixed workplace is empowering to women without anyone needing to make a big deal out of it.

Sex differences exist, of course they do – but they are biological, and as such they don't affect intelligence or aptitude. Gender differences are a social construct and as such they manifest differently at different times in history. Nobody now would claim, as Victorian physicians did (they were following the science, of course), that a disease called anorexia scholastica affected women, and only women, who studied maths.

Ada the beacon is also Ada the lighthouse; the warning of rocks ahead. There are plenty of obstacles for a woman who chooses a STEM career – assuming she ever gets that far – and if she is better than good she will have to prove it many times over.

Even after she is dead.

*

Babbage, for all his genius, was not unlikely. He had money and education, position and patriarchy. As a young student, his parents didn't get letters home, worrying that too much study might drive him mad, or weaken his vital organs.

Babbage belonged to his world. His fights (many) were fights among equals. Men are allowed to disagree violently with one another without disrespecting one another. Read any account of Babbage, favourable or not, and the tone is respectful. Read any account from the Ada-deniers and the dominant, monotonous tone is lack of respect. In fact, it is contempt.

Ada was unlikely. True, she had money, and that helps a lot, but she was a woman in a man's world.

She was legally barred from going to university, or earning a degree, let alone securing a prestigious post as a Cambridge professor. She couldn't get a job in engineering, or work with Babbage's great friend Isambard Kingdom Brunel, who was often to be found with Babbage poring over new calculations for his railways or steamships.

Ada wasn't free to go out and build the world. She was the legal property first of her father, and then of her husband. When she married at 19, she spent the next 3 years having 3 children. Babbage's wife did that job for him.

Ada had no social structure in which to exist as Ada. For some reason, maybe the Byronic blood, she didn't care. She was lucky to have a husband who bored her, but who let her do as she liked – or who didn't notice what she was up to most of the time, and that included many affairs, and a great deal of money lost on horse racing.

She was freakishly lucky to have been given a maths tutor, and blessed to find that she had an intuitive understanding of numbers, as well as the necessary application for long hours of solitary study, at a time when women of her class and kind were encouraged to be frivolous man-pleasers – at least until they lost their looks, at which point they could devote themselves to charitable causes.

The probability of there being such a person as Ada Byron, later Countess of Lovelace, is vanishingly small. But there she is, in history just at the right time to meet Charles Babbage, and to design the first software programme for the first (never built) computer.

It all starts with a ground-breaking paper Ada wrote – without actually realising she had written it, because back in the day women didn't write scientific papers.

It happened like this …

Babbage had gone to Turin in 1840 to give a lecture on his Analytical Engine. The only person to take any notes was an Italian engineer called Luigi Menabrea. He published these notes a couple of years later in French, in a French journal. Then, as now, Europeans are likely to speak more than one language, while the English don't bother. In the sunny uplands of Brexit, this legacy of Empire will have to change.

Ada, though, was female, and had the usual accomplishments of an upper-class woman destined to entertain Important Men at Dinner Parties. She was fluent in French. She decided to translate the Menabrea paper.

As she translated the straightforward account of how the Analytical Engine worked, she added footnotes of her own. These footnotes are nearly 3 times as long as the original paper by Menabrea. In her notes she also details the first full-length 'programme' for the machine, and she separates out the functions of what we now call hardware and software.

Not stopping there, she understood that if the machine could be programmed to calculate *something*, it could be programmed to calculate *anything*:

> The bounds of arithmetic were outstepped the moment the idea of applying the (punched) cards had occurred, and the Analytical

Engine does not occupy common ground with mere calculating machines. It holds a position wholly its own; and the considerations it suggests are most interesting in their nature. In enabling mechanisms to combine together general symbols in successions of unlimited variety and extent, a uniting link is established between the operations of matter and the abstract mental processes of the most abstract branch of mathematical science.

In other words, if the machine could manipulate numbers (arithmetic), it could also manipulate symbols (algebra).

Computing uses symbolic logic – a development on basic algebra that was a new discovery in Ada's day. Ada's maths teacher, Augustus De Morgan, was a pioneer in the field, corresponding with the self-taught mathematician George Boole (1815–1864). It is fair to say that there isn't a piece of technology on the planet that is operating right this second without using Boolean logic. It's the basis of computing. And it's down to True or False.

Boole proposed what he called 3 operators: AND. OR. NOT. He demonstrated that all logical relations (at least algebraic ones – and perhaps even your life plan) could be expressed using his laws, or rules. Boole believed that human thought could be reduced to a series of mathematical rules – and 'reduced' here isn't reductionist in a negative way: Boole was looking for simplification and clarification. To achieve this goal, he used binary numbers – zeros and ones, where 0 = False and 1 = True, and the awesomely impressive idea of Truth Tables. Yes, the probability of any answer being true or false could be determined via Truth Tables. Which is cheering for over-anxious Humanities types.

How does it work?

Well, in arithmetic we deal with specific numbers, known as fixed values, such as $2 + 2 = 4$. In elementary algebra we are dealing with quantities without fixed values, called variables. X or Y do not necessarily mean 2 or 4. The values of X and Y will change – they are variables. But whether it's arithmetic or

:38   21/05/2024

Receipt for Returned

atron details:

ame ---------------- P********.*
                                    ***a

D) ----------------- 2***********
                                     *1

Outstanding fees ----------- £0.00

The Topeka school
Item ID: 30129088500102

Item count: 1
Successfully returned: 1

elementary algebra, our working operations are multiplication, subtraction, and addition. In Boolean algebra we're not doing multiplication, subtraction or addition, we're doing Conjuction (AND), Disjunction (OR) and Negation (NOT). This is because the variables are not numbers, but truth values – expressed as True or False (1 and 0).

To give a quick example:

My garden X is made up of soil Y and plants Z, so if I am using elementary algebra, then my garden is Y (volume of soil) + Z (number of plants) = X (garden).

Elementary algebra would set a number value to the garden. How many plants, what quantity of soil, etc?

In Boolean algebra, I would end up with a 'truth'. Is it a garden or not (according to my parameters of what is a garden)? So, if there is NOT soil and NOT plants, it isn't a garden. If there is NOT soil and 1 plant, it isn't a garden. If there is soil, but NOT plants it isn't a garden. But if there is soil AND plants – then it is a garden. If Z AND Y are 'True' then X is 'True')

Then we have fun adding in W = weeds and C = Concrete to decide what a garden 'is'.

More everyday – and for non-gardeners out there – if I am searching online for singer Keith Urban, the search engine will use Boolean logic, arriving at Keith AND Urban to be 'True'. Keith Richards or Urban Escapes would be eliminated by the Boolean search engine as 'False'.

There is no need for any human to programme ahead all the possibilities of 'Keith' or 'Urban'. Boolean logic allows the computer to work through all the variables (possibilities) and come up with the answer by itself.

Ada writes:

> I want to put something in my notes as an example of how an explicit function may be worked out by the engine without having been worked out by human heads and human hands first.

*

This was a long way past logarithm tables. Ada was heading for the stars.

In the 1840s, computing terms were not in general use, so Ada explains what she means by an operation: 'By the word operation we mean any process which alters the mutual relation of two or more things. This is the most general definition and would include all subjects in the universe.'

All subjects in the universe. That is pretty far-out.

Ada was amused to speculate that the Analytical Machine could easily be programmed to write music. (Babbage was waging a one-man war on street musicians outside his house – especially those who had a barrel organ.) Ada, to cheer him up or madden him, *who knows?* wrote: 'The engine might compose elaborate and scientific pieces of music of any degree of complexity or extent.'

Fun or not, to get a sense of how ahead Ada was in her thinking, we have to go forward in time 140 years, to the 1980s.

American classical composer David Cope had plenty of success; his work was performed at Carnegie Hall, and he was able to make a living, as well as being critically successful.

During the mid-1970s, an unexpected period of creative block had caused him to start experimenting with another intelligence as well as his own. Had it been a human collaboration no one would have demurred. But his collaborator was AI.

Cope began inventing and coding his own music-making programmes. The programmes unpicked the structures behind Mozart, Vivaldi, Bach, and began to recombine what existed already into what did not. By the 1980s, Cope could go out to buy a sandwich for lunch and come home to 5,000 'new' Bach chorales.

Cope was excited about what he was discovering. Critics, though, were unimpressed – 'no soul', 'rigid', 'formulaic' were the usual responses. Except that audiences asked to guess what pieces of music were 'real' (i.e. human) often ticked the computer music as the real thing – the real Bach, the real Vivaldi.

Cope is a thoughtful human, and for him, the interest of creating programmes that create music is as much for the questions it asks us, the 'real' humans, to consider. Big Question: What is 'real'? Next Big Question: What is 'creativity'?

40 years on from Cope's first experiments, AI has come a long way, and contemporary music is making good use of AI programmes to generate music. IBM's Watson Beat, Sony's Flow Machines, Spotify's Creator Technology, plus the user-friendly Amper that offers 'musician-trained Creative AI'.

The process of training the AI is exactly as Ada envisioned it nearly 200 years ago. (Go girl!) The programmer-human feeds in as much existing music material as possible so that the programme itself can analyse patterns – tempo, beat, chord progressions, vocals, length, variations – then the human composer sets the parameters – upbeat anthem, moody love song – and works creatively with what comes out.

Here's what Amper says about its tools:

> Amper Score™ enables enterprise teams to compose custom music in seconds and reclaim the time spent searching through stock music. Whether you need music for a video, podcast, or another project, Score's Creative AI quickly makes music that fits the exact style, length, and structure you want.

And just a quick recap of what Ada said in 1840:

> The engine might compose elaborate and scientific pieces of music of any degree of complexity or extent.

*

Ada didn't believe that programme music would or could be 'original', which wasn't the same thing as saying it wouldn't be entertaining or satisfying. Ada was adamant that a computer could analyse data, but a computer could never think for itself:

> The Analytical Engine has no pretensions to originate anything. It can do whatever we know how to order it to perform.

Ada's view on the kind of music a programme might compose is pretty much in line with where we are right now with AI.

If you want background music, hotel-lobby music, something noisy and shiny, or soulful, let's say for an advert or a promo, then AI can do it for you with little or no tinkering around by the human – and, as Amper points out, no royalties will be payable! This is music as a commodity.

Working with the AI to make 'original' music is still an area of dispute. Many musicians say it is for lazy people who can't be bothered to learn to read music or to play an instrument, and who want the hard work done for them, while raking in the fame and fortune.

Brian Eno believes that the excitement is in the collaboration – a skilled musician working with AI. Eno released his 26th album, *Reflection* (2017) as a CD but also as a generative app that recombines and changes the original music depending on the time of day. Eno said, 'My original intention was to make endless music … like sitting by a river: it's always the same river, but it's always changing.'

Collaboration, though, doesn't always end up with the human part of the experiment.

Hatsune Miku is a Vocaloid software voicebank. She was released in 2007 by the Japanese company Crypton Future Media. She is always 16. She is 5'2" tall – and weighs an anorexic 42 kilos. Turquoise is her colour. She sings, appears in video games, and tours as a hologram. Her huge popularity in Japan seems in part to come from Shintoism – where inanimate objects are believed to have souls. Judging from her fanbase comments, Miku is a believable soul. She's not exactly inanimate – but she isn't human either. She raises the question: in what sense does it matter?

Does it matter if the wind makes the music, blowing through bells and pipes? Or if a piece of software makes the music? Or if Bach does? Or if you do?

I am not suggesting equivalence of quality – and that is a different debate – only whether or not the who or the what is misleading.

Set against that thought is the fact that the biggest touring bands in the world are still old-fashioned (and increasingly old-aged) guys who write their music and play their instruments. But this is probably the end of an era.

As David Cope puts it: 'The question isn't whether computers possess a soul but if we possess one.'

Alan Turing, the British mathematician who designed and built the Enigma code-breaking machine at Bletchley Park (Turing was played by Benedict Cumberbatch in the movie *The Imitation Game*), wasn't interested in whether or not a computer could have, or would have, a soul, but he was interested in whether or not a computer could originate (as well as learn) independently of human input.

Turing took issue with Ada Lovelace on this matter – though again we have to remember that he took issue 110 years after her statement that a computer could not *originate* anything (not the same thing as arriving at the right answer by itself, by means of Boolean programming – Ada was thinking about the leap-capacity of human intelligence, something she didn't believe a computer would ever demonstrate). Ada's limits, said Turing, were a product of her time. She was working with what she had – and what she had was nothing … because Babbage didn't build it.

Ada, though, was good at working with what she didn't have – she didn't have male privilege, she didn't have a formal education, she didn't have a computer. She was Ada.

Turing dusted Ada down and brought her back from the dead in 1950, when, after carefully reading her work, he responded to what he called Lady Lovelace's Objection – that computers cannot *originate*. In this they are not human or like a human.

His answer to Ada – their conversation across time – was the Turing Test.

The 1950 Turing Test is a test of a machine's ability to appear to humans as equivalent to, or indistinguishable from, a human.

Google claims that their voice technology assistant Google Duplex has already passed the test – at least when it comes to booking appointments for you by phone. Fooling the human receptionist on the other end of the line counts as a pass – and Google Duplex has achieved that, with tone modulation, word elongation, and 'thought' pauses that all sound like a realistic human. But those responses have to be programmed – so Ada is still correct.

Turing thought that machine intelligence would pass his test by the year 2000. That hasn't happened, but we are getting closer – and while we are getting closer we might decide, or AI might decide, that it doesn't matter.

Mary Shelley may be closer to the world that is to come than either Ada Lovelace or Alan Turing. A new kind of life-form may not need to be like a human at all (the cute helper bot or the virtual digital assistant may be just a distraction, a sideline, a bridge. Pure intelligence will be *other*) – and that's something that is achingly, heartbreakingly, clear in *Frankenstein*. The monster is initially designed to be 'like' us. He isn't and can't be. Is that a message we need to hear?

*

While amusing herself about Babbage's machine being able to make music of a kind (the kind that would annoy Babbage), Ada never assumed it would write poetry.

BotPoet (just Google it) was not available in the 1840s. (Clue: it's a Turing Test for poetry.) Ada was the daughter of Lord Byron, so it was important to show some respect to poets. For the English, ours being the Land of Shakespeare ('This precious stone set in the silver sea'), poetry was as close to God as you could get, apart from mathematics, and most people couldn't understand maths. And you certainly couldn't recite equations at Important Events.

Ada described her own work as poetical science; therefore it was doubly creative and not about to be reproduced by an algorithm.

As of now, computers aren't that great at poetry. I suspect because there isn't a clearly analysable form – I don't mean formula. We can, and should, learn how a poem works, because that increases our pleasure in it, but the emotional hit is odd and harder to pin down.

The hit is like the ghost in the machine – we can see how the poem is made, but the elusive sprite-like thing refuses to get in the bottle and, well, be bottled. You/AI can learn how to write a sonnet or a villanelle, or in couplets, or in blank verse, but how to make the magic happen? The mysterious part of poetry is like an emergent property; it happens out of language, but it is not exactly in the words, or even the order of words – just as consciousness arises out of brain/mind but isn't brain/mind. And yet – what else is there in poetry but words? What else is there in consciousness but brain/mind? Answer: something else. Weird but true. (See the essay 'He Ain't Heavy, He's My Buddha'.)

*

Fiction, though – well, that's a different thing. Everyone has a novel in them, don't they?

There are plenty of novel-writing apps online now – to organise your story, streamline your plot, twist it when it needs twisting, manage dialogue, widen (or not) your vocabulary, and do the simple stuff like grammar. novelWriter will take your garbled strapline and offer you a packaged version of how it might work out.

For instance, if I type in: *Cat falls down mineshaft and discovers a secret world of giant mice with computing skills,* novelWriter will help with all the components of character (probably a lot of characters if they are mice), and plot twists – and yippee, I have written a novel (about mice).

On the other hand, if I put in: *Young man in the reign of Elizabeth the First wakes up one morning in Turkey as a woman,* I probably won't write Virginia Woolf's *Orlando.* Anyway, it's been done.

Perhaps mathematician Marcus du Sautoy, author of *The Creativity Code* (2019), is correct when he says that the 2050 Nobel Prize for Literature will go to Alexa. Or perhaps that's the Revenge of the Maths-Head.

Self-generated novels. Nothing new. The satirical magazine *Punch* offered this Babbage spoof back in 1844:

> Sir … I have been completely successful in the production of a New Patent Mechanical Novel Writer – adapted to all styles and all subjects…
>
> Babbage

Followed by testimonials:

> By its assistance I am now able to complete a novel of 3 vols in the short space of 48 hours, whereas before, at least a fortnight's labour was requisite for that purpose …

And:

> Lord W has now nothing more to do than throw in some dozen of the most popular works of the day, and in a comparatively short space of time draw forth a spick-and-span new and original novel ...

No more labouring lone genius putting in the 10,000 hours of slog. This is pop art before Warhol. This isn't art for the masses, this is art from the masses, and masses of it.

Not rare. Not strange. Not special. Continuous. Combinatory. Computer. (Those 5,000 extra Bach chorales done over a lunchtime sandwich. Brian Eno's forever app.)

What will this mean for humans? For creativity?

Or do I mean, what is meaning? For humans? For creativity? We shall have to reimagine those terms: Humans. Creativity. Meaning.

Here's Tim Berners-Lee, father of the World Wide Web:

> What matters is in the connections. It isn't the letters. It's the way they're strung together into words. It isn't the words, it's the way they're strung together into phrases. It isn't the phrases, it's the way they're strung together in the document ... In an extreme view the world can be seen as only connections, nothing else.
>
> *Weaving the Web*, TB-L, 2000

Or, as E. M. Forster put it in 1910, in his novel *Howards End:* 'Only connect.'

Or, as Ada put it, 'A uniting link is established between the operations of matter and the abstract mental processes ...'

In 2021 Google's next goal is ambient computing: that's wraparound connectivity. Hardware/software/user experience/machine-human interaction. The Internet of Things. All integrated. From cat-flaps to coffee machines. Voice-activated digital assistants. 3-D printers. Smart homes. All working together. Invisibly. Permanently. No need for clicks. Think and it will appear. Magic lamp. Operations of matter. Abstract mental processes.

Ultimately – and Ada was right about this – the uniting link between the operations of matter and abstract mental processes is to reimagine – completely – what we call 'real'. This reimagined 'real' will soon be what we call the world.

A Loom with a View

The brain is an enchanted loom where millions of flying shuttles weave dissolving patterns.

Charles Sherrington, neurophysiologist and
Nobel laureate, 1940

To understand the new journeys and destinations opening up to the human race, via emerging and developing AI, it's helpful to think about how we arrived where we are now. The huge decisive changes in our world are the changes that began with industrialisation.

When we consider that the earth is about 4.5 billion years old, and that the oldest Homo sapiens fossil, discovered in Morocco in 2017, is 300,000 years, then what has happened in the last 250 years is time on a tiny scale – my 1780s house in London has lived that long.

Not the next 250 years but the next 25 years will take us into a world where intelligent machines and non-embodied AI are as much a part of everyday life as humans are. Many of the separate strings we are developing now – the Internet of Things, block-chain, genomics, 3D printing, VR, smart homes, smart fabrics, smart implants, driverless cars, voice-activated AI assistants – will work together. Google calls it ambient computing: it's all around you. It's inside you. This future isn't about tools or operating systems; the future is about *co-operating* systems.

The technology is moving fast. The Data Age will be the biggest change yet seen on Planet Earth – bigger than the first Industrial Revolution – when we came blinking out of a centuries-old rural economy, and wide-awake into the daily nightmare of the industrial economy. It's not helpful to write Machine Age = Progress. One word isn't enough.

The social, psychological, and environmental costs of progress have been and are immense. Accounting is more than numbers; accounting is accountability. This time, the real costs of progress have to be faced and factored in.

Earth scientist James Lovelock (creator of the Gaia theory) says that humankind is at the end of what he calls the Anthropocene (from *anthropos*, Greek for 'human being') and about to begin a wholly new departure, into the Novacene. (In astronomy, a nova is a star that suddenly releases a huge burst of energy.)

Like Ray Kurzweil (creator of the Singularity theory), James Lovelock believes that this departure will be irreversible. Artificial Intelligence will not be a tool for long; it will be a life-form.

But before that happens, all of us humans will be living in computer-world; I don't mean a simulation – though that could easily be true already – I mean societies inseparable from, and dependent on, our interface with AI.

We need to learn from the past. That is what the past is for.

So, let's take a trip back to where the future began. In Britain. Steam power. The steam-powered loom.

Spinning and weaving are two of the oldest skills in human history – going back in time at least 12,000 years. Humans need clothes and coverings. What changed the way we made those goods was automation.

*

Britain in the 1700s was all about wool. Think *sheep*.

Where I come from, in Lancashire, sheep used to be known as white gold. You could wear them, make rugs out of them, and eat them. Unlike cattle, sheep are modest in their wants, and can manage harsh winters outdoors. Sheep were the pre-industrial economic miracle.

The British were obsessed with wool. Anyone not dressed from head to foot in woollen broadcloth and knitted socks was letting the side down. Even in August.

Cotton doesn't grow in western Europe. Cotton goods imported to Britain in the 1700s came from India, so money spent on Indian cotton stuffs was money not spent on British wool.

Cotton, though, is floaty, light, lovely to dye, quick to wash and easy to dry (consider a wool suit dripping over an open fire). Above all, when worn next to the skin, cotton is not scratchy. It is no wonder that women didn't wear knickers until the 1800s.

But cotton was expensive. In 1700, a skilled spinster with a pedal-powered wheel took 40 hours to turn a messy heap of cotton fibre into a pound (450 grams) of yarn.

Women were in charge of this work. Women were the original spinsters.

To be a spinster was a proper job title with cash and kudos attached. A spinster wasn't a past-it woman who had 'failed' to find a man, she was a valued member of the community, who could, if she wished, support herself independently.

Twisting wisps of fibre onto wheels is the kind of process that lends itself to mechanisation. Whatever is repetitive can be done faster by a machine. After thousands of years of women and men spinning by hand, James Hargreaves' Spinning Jenny (1764) and Samuel Crompton's Spinning Mule (1779) had the process of spinning a pound of yarn down from 40 hours to 3 hours – and it was soon cut to 90 minutes. Then in 1785 Edmund Cartwright did for weaving what the other boys had done for spinning. The power loom went into the new factories and by the turn of the century the great age of industrialisation had begun.

The Industrial Revolution was a practical revolution.

Humans reworked everything we had learned over millennia of trial and error – clothing, manufacture, transport, heating, lighting, weaponry, medicine, construction.

Faster. Cheaper. More of it.

Smaller takes a while to get into the picture because smaller isn't important until the 1950s electronics revolution that is transistors.

For the 19th century the slogan is: Bigger is Better (top hats, crinolines, chimneys, iron bridges, engines, ships, cannon, and factories of course. Vast, non-human in scale. The machine as monster – prefigured by *Frankenstein*).

> The steam-engine had no precedent. The mule and the power-loom entered on no prepared heritage. They sprang into existence like Minerva from the brain of Jupiter.
>
> *Notes of a Tour in the Manufacturing Districts of Lancashire*, W. Cooke Taylor, 1842

By 1860 Britain was still the only fully industrialised economy in the world, producing HALF of the world's iron and textiles (just dwell on that for a second).

Imagine the new cities. The massive steam-powered, gaslit factories. The back-to-back tenement houses thrown up in between. The filth, the smoke, the stink of dye and ammonia, of sulphur and burning coal. The ceaseless activity day and night, the deafening noise of looms, of coal delivery, of waggons on cobblestones, of relentless machine clatter. Of teeming human life.

Manchester, where I was born, soon became the cotton capital of the world, and until World War One, 65% of the world's cotton was processed in Manchester, aka Cottonopolis.

America supplied most of that raw cotton. Britain's relationship with its former colony was crucial. Millions of acres and hundreds of thousands of slaves grew cotton. In 1790 the southern state plantations had been exporting around 3,000 bales. By 1860 they were exporting 4.5 million bales.

The Industrial Revolution is a tipping point in environmental history. It's when fossil fuels come out of the ground in world-changing quantities. Britain had a lot of coal and used it to gain an early advantage. Coal burns hotter and for longer than wood and, most importantly, the temperature can be kept steady once achieved. A coal-fired furnace produces an impressive head of steam.

Turning heat into movement is what steam power made possible – first with water pumps down the mines (Newcomen engine), then with steam-powered looms and, later, railway engines and iron ships. Unprecedented. A total break with the past. What could be less intuitive or less natural than a ship (a vessel that needs to float) made out of iron? A material so heavy it sinks. No oars. No sails. No need of wind. Powered by a kind of magic. Filthy, dirty, smelly black magic.

This was not just a new kind of technology; it was a new kind of filth. It is the beginning of pollution – of land, air, water, crops, and humans.

Friedrich Engels watched the money and the misery pile up in equal measure, though by no means equally distributed. In 1845 he published *The Condition of the Working Class in England*.

> A horde of ragged women and children, as filthy as the swine that thrives upon the garbage heaps and puddles – neither drains nor pavements – standing pools in all directions – the dark smoke of factory chimneys. A measureless filth and stench.

In accounts of the period, the repeating words are: *Filth. Stench. Noise.*

Karl Marx, born in 1818 (the year Mary Shelley published *Frankenstein*), walked the streets of Manchester with his friend Friedrich Engels, and formed from his experiences there the basis of *The Communist Manifesto* (1848).

The new machinery in the factories had increased output by 25% per person (don't fly over that statistic – consider its implications). Wages, though, had scarcely risen by 5% from their pre-industrial levels. And now people were out of the countryside, with its free food and fuel economy, and living in the burgeoning cities – where everything had to be paid for with money.

Conditions were foul. No running water. No sanitation. Pigsty dwellings. Filthy air. Life expectancy was around 30 years for the slum-housed factory workers. So much money. So little equality. Folks called Manchester the Golden Sewer.

Marx, writing in chapter one of *The Communist Manifesto*, had this to say, and it sounds eerily like the 21st-century 'move fast and break things' (Mark Zuckerberg – Facebook) mentality we associate with Big Tech:

> Constant revolutionising of production, uninterrupted disturbance of all social conditions, everlasting uncertainty and agitation distinguish the bourgeois epoch from all earlier ones ... All that is solid melts into air.

For Marx, on the ground at the time, technological progress was the guarantee of social revolution, because progress was being driven without any regard for the human cost of the soaring profits. Men and women were paying with their lives – and it wasn't the sudden loss of life expected in wartime, but a long-drawn-out hell on earth. The people's revolution would happen, said Marx, because the alternative was unliveable.

Even now – or, rather, again now – because of current neo-liberal propaganda, the convenient and simplistic reading of the Industrial Revolution is progress for all. Just a few teething troubles along the way.

The word Luddite still means an old-fashioned type who is anti-progress. But the Luddites of the early 19th century were not against progress; they were against exploitation.

The 1812 Frame-Breaking Act made damage to a power loom punishable by death. Luddites were risking their lives, not to return to some sentimental rural Arcadia, but to earn enough to feed their children.

In the new world order, capital and property were protected by the death penalty. Human labour was left unprotected. The working man and woman had to manage low pay, hateful conditions, and insecurity of employment. For them, there was neither progress nor improvement.

The physical and mental anguish of the first 50 years of the Industrial Revolution was increased by the rise in population – from around 10 million in Britain in 1800 to almost double that by 1840. More children meant more mouths to feed – but children are more than mouths. They are *children*.

Only in 1832 were children under 9 years old barred from factory work. Only then were children of 10 years and older limited to a 48-hour working week.

Shall I write that again? A 48-hour working week for 10-year olds.

The fact is that the first 50 years of the Industrial Revolution were years of unprecedented misery as well as unprecedented innovation. And it wasn't just factory hell.

Until 1815 Britain was intermittently at war with France – largely as a counter-revolutionary precaution. The success of the French Revolution in 1789, with its rallying cry of Liberty, Fraternity, Equality, promised – or threatened, depending upon which side you were on – a new kind of social order. And it was only a few years after the American Declaration of Independence in 1776 – and its famous opening: WE THE PEOPLE.

In England, following these gigantic political upheavals, and with a view to mobilising the British out of their hierarchical class-mentality, Thomas Paine, an American living in England, published *The Rights of Man* (1791).

Paine's book identifies and explores what we think of as a modern concept – and one that is now under threat: the idea of 'civil rights'.

Paine says that people should be able to expect security – of their persons, and of what they own or have access to (like common land), of jobs (job security) and of wages. Tax should be progressive – the rich pay more than the poor. Education should be available and subsidised. There should be universal old-age pensions. Maternity pay. Funded apprenticeships to aid young people.

Governments, said Paine, exist for the good of all, or they are tyrannies and should be overthrown.

Britain's wars with France created an enemy elsewhere, a convenient, sure-fire distraction from poverty, exploitation and inequality at home.

Wars are also an opportunity to draft men into the military. A consequence of the absence of men is that women must do their work. Women and children were preferred for factory work because they could be paid less.

Paying less (then as now) was gender-based, with the added advantage of de-skilling a job.

Whenever a woman took on a job previously done by a man, the job itself was downgraded. What looks like male chauvinism in the Industrial Revolution, as working men refused to train women to work the machines, was really a fight for survival. A man knew that if his wife or his sister was able to do his job, then he would be paid less. And in lean times only one of them would be offered work. Not him.

Not everyone, though, was in factory work. In the early years of the Industrial Revolution, the rural economy was still pre-eminent.

This period of British history is the history of enclosure; public or common land seized by local landowners. I have no idea why history books still use the word 'enclosure'. Sounds like a hug.

Enclosure was a land-grab by the rich for the rich. Enclosure was theft. It was colonialism on a domestic scale.

In 1801 a General Enclosure Act made land-grab easier for the grabbers and much harder for anyone else – even modestly prosperous small farmers – to do anything about. Compensation was paid to some people, but money is soon spent, whereas land remains.

Enclosure of common land, or peppercorn-rented land, left nowhere for the cottager or terraced-house dweller to keep a few sheep or a cow, or to grow extra crops beyond the garden. Wood could not be gathered freely to heat the home, or for cooking. What had been held in common for all became private property.

Of course, this has always happened, and all over the world. What was different this time, in England, was its systematic long-term enforcement by Acts of Parliament. From the early 1800s to the late 1870s, enclosure acts favoured the large landowner.

Enclosure had a catastrophic effect on the individual and local economy – often balanced on an informal system of exchange. This literal lack of land, and the extras it provided in terms of food and fuel, drove many more people through the dirty doors of the factory system than would have happened otherwise – and I say this because accounts of the period that are accounts of the workers' own situation, and

not some self-interested whitewash, show how much the
factory system was hated by those made to work in it.
(Anyone wanting to read further on this could start with
Engels, and also E. P. Thompson's *The Making of the English
Working Class.*)

Forced off the land by enclosure, an over-supply of impover-
ished workers drove down factory wages.

Before the Industrial Revolution, money in your pocket
mattered less. If you can put food on the table from your own
veg patch, from your hens, pigs, cow, and sheep, and if you can
make the clothes you need, gather fuel, and sell or barter enough
for your other needs, you may have very little money in your
pocket, but be living a decent life.

Pre-Industrial Revolution, even the wealthy might have
extensive land but very little cash – that's what marrying an
heiress was for.

The assumption that people are 'lifted' out of 'poverty' by
charting wage increases says nothing about the poverty they
have been plunged into by economic 'progress'.

Wages have only been used as a metric of progress since
industrialisation and the creation of a wage-earning working
class.

And, of course, wages say nothing about the well-being,
independence, happiness, and mental health that come from
productive, pleasurable activities that are economically viable,
but resistant to straight-line capitalist metrics.

Wages, though, at last (and we are talking about 40 years –
longer than an average lifespan back then) did start to rise.

The world's first trades union organised itself in Manchester, as workers realised that their power could be stronger in
negotiation than in outright revolution.

As boom-time boomed on, it was in everyone's interests to
pay factory workers more money, so that they could spend it on

goods, as well as drink. Gin addiction was the British vice – and had been before industrialisation – but only in cities – now the lethal combination of new cities and new slums created a drunken underclass on a grand scale.

14-hour days, potato soup, no running water, and 10 of you sleeping on a dirt floor – with no hope of change?

Pass the gin.

Marx was right: the old loyalties, the old traditions, and any sense of paternalism had been destroyed by the brutality and betrayals of the twin hits of enclosure and industrialisation. It was nearly a knock-out blow. Reeling back to their senses, workers managed, somehow, to initiate collective action, and with it a growing sense of community. (Solidarity.)

It wasn't the people's revolution Marx had predicted. It was, though, the formation of a whole new class of people, not bound together, as of old, by guild, or craft, or parish or family ties, or religious denomination, but by the single simple fact that they were workers.

And this extended past local workers at home.

In 1862 at the Free Trade Hall in Manchester, Lancashire cotton workers voted against supporting the American south and its slave interests and resolved to support the north and Abolition. The workers knew this would involve considerable hardship (and they knew what considerable hardship involved). Yet by now globalisation meant something more than free trade; it was the beginning of the Information Age.

Workers in Manchester were keenly interested in what was happening to Black slaves in the Southern states of America. Most people in Manchester had never seen a person of colour. There were Black servants – especially in London – but the UK was overwhelmingly a white place to be.

The solidarity of cotton workers, some on American planta-tions, some in English factories, divided by the Atlantic, transcended jingoism, nativism, and racism. This is not to say that racism wasn't present and ugly once there was noticeable immigration to the UK (I date this after the 1948 British Nationality Act), but the support of the UK cotton workers against US slavery was real.

In 1865 the American Civil War put a legal end to slavery. The subsequent 100 years in America were the years of the slow journey to do more than end slavery. Segregation itself had to be ended before Black Americans could win full equality in society and under the law.

The legal end of slavery, imperfect and unfinished as it was, is such an important moral milestone. The legal end of slavery marked the end of any claim to the 'natural' right of one person to own another.

This went straight to the heart of the woman question.

All women, whatever colour, whatever class, were the legal property of their nearest male relative. Widows had some autonomy.

But why should a woman be the legal property of her male relatives? Why should whatever she owned, and whatever she earned, belong to her husband automatically, and by right? A woman's children were not her own. They were the property of her husband, until the male child came of age, and the female child passed as a chattel to another male, in marriage.

Why should a woman not expect equality under the law, at work, in society, in the family unit?

> The legal subordination of one sex to the other is wrong in itself and now one of the chief hindrances to human improvement. No slave is a slave to the same lengths and in so full a sense of the word, as a wife.
>
> John Stuart Mill, 'The Subjection of Women', 1869

Mary Wollstonecraft (the mother of Mary Shelley), had written her own version of Tom Paine's *Rights of Man. A Vindication of the Rights of Woman* argued for education, voting rights, property rights and employment for women. It was even less well received than Paine's, by those all-male elites for whom social democracy was tantamount to revolution.

But *Rights of Woman* was the starting gun for feminism.

Industrialisation pulled the trigger.

In the factories, and in the vast, expanding towns and cities, large groups of women were meeting one another in workplaces, and on the streets, in a way that was wholly new, both in scale and dailiness. Men have always met outside of the home. For women, the factory experience, brutal though it was, brought them together beyond anything that family, village, farm, church, or domestic service had made possible.

These women were talking to each other. These women could see that they were working as hard as men. They could see that they were also running the home (hovel), and managing the children. How could a woman be classed under the law as a minor, with no legal rights, when she was working 60 hours a week and holding a family together? Why was she paid less than a man for the same work? Why was she the legal property of her nearest male relative?

It was not until the 1970s, in Britain, that sex-based discrimination was made illegal. That included equal pay, equal access to credit and mortgages, equal opportunity in education and in work.

In 1974 the Equal Credit Opportunity Act became law in the USA.

If we take this in the context of the timeline of economic and social progress (the buzzword), we're looking at 200 years since old Jimmy Hargreaves called his world-changing new invention Jenny (Spinning).

Women are still on the road towards full equality with men. Across the world millions of women are still on a track going nowhere. Progress hasn't materialised for them. Around

800 million people in the world are illiterate. Two-thirds of them are women – and that percentage hasn't changed for 20 years.

Progress.

What do we mean by the word progress?

Technical innovation? Social change? Living standards? Education? Equality? Globalisation? For everyone?

Lessons from the past show that governments need to legislate in order for innovation to benefit the many and not the few.

In the 19th century we start to see legislation working to enact what we took for granted in the 20th century:

Factory Acts to limit working hours and to mandate paid holidays.

Rudimentary health and safety in factories and the workplace.

Education for children. The idea of childhood as a protected period in the life of a human being.

Unionisation to provide job security and a living wage.

Sanitation (drains, sewers, running water) and street-lighting paid for by taxing corporations. Slum-dwelling legislation.

Affordable public transport for workers.

Libraries for working people. Night school.

Even city parks. Yes, parkland was the 19th-century way of saying – Oops, we did steal all the common land during Enclosure, but never mind, here is some grass and a fountain, etc (and often a few statues of 'great' men), all provided, without charge, for the mental health and moral uplift of the working class.

Working class. As the great historian of labour E. P. Thompson put it, 'Class is a relationship. Not a thing.'

That is, class is mistaken for a noun, like horse, or house, but class doesn't exist in its own right. It isn't a thing. And it isn't a natural phenomenon – like gravity. In an equal society, class would not exist. Social division is relational – it isn't a pre-existing condition.

The 19th century laissez-faire opposition to workers' rights, and to a social contract, was soundly defeated after World War Two, both in America and in most of Europe.

Keynesian stimulus, backed by America's Marshall Plan – the plan that lent money to Europe to buy American goods – bucked the budget-balancing ideology that was really about keeping some people poor in order to make other people rich. Spraying money at the problems of poverty allowed shattered economies to grow.

Between 1945 and 1978, the US economy more than doubled. Germany rebuilt herself as Europe's most efficient economy. Britain developed a welfare state, the NHS, and a housing programme. Of course, it wasn't perfect – nothing is perfect when humans are involved. Yet when President Kennedy said, in 1963, that 'a rising tide lifts all boats', he was speaking the truth.

The factory system was – at last(!) – paying top wages, both in spite of and because of automation. The Ford car plants in the USA had innovated the factory system with the production line and robotics, but the money was flowing fast, and the future looked bright – not just for a few people in the top slice, but for everyday folks.

Capitalism is adaptable. The Marxists were wrong about that. Capitalism isn't a rigid system. Capitalism can work alongside socialism. Judging from what has happened in China, Russia, and the former East Germany, capitalism can manage communism too.

That has to be admired and I do admire it.

Capitalism is Darwinian in the true sense; not the cheap line about 'survival of the fittest', but the ability to adapt to different and surprising circumstances – and stay the course.

Socialism need not position itself as the polar opposite of capitalism. Socialism can temper the excesses, challenge the bull-headed free-market mantra. The market is not God. And, while markets may correct their own excesses in the long run, as Maynard Keynes remarked, 'in the long run we are all dead.'

I was a teenager in the 1970s, as the post-war social contract in Europe and America began to break down.

There were so many contributing factors: low productivity, inflation, the oil crisis, the 3-day week, the aftermath of the Vietnam war. Nixon's unilateral abandonment of the gold standard in 1971. Absolute mayhem with unionised labour. (In Britain it was the miners looking for a 35% pay increase that triggered the 3-day week. Margaret Thatcher would smash them for that when she was in power in 1985.)

Looking back, it seems now that the 1970s were really the end-time years of the West's Industrial Revolution. We were waiting for the Computer Age that wasn't quite ready. Desktop computers didn't appear till the mid-1970s, and they weren't built by companies, but by guys in garages.

Acceleration also leads to exhaustion – because humans aren't a version of Moore's Law, where the number of transistors on a microchip doubles every 2 years, increasing speed and lowering price. Humans sometimes need to slow down. We run out of ideas.

The late 1970s feel to me like an energy-lag on the Left. There was no viable new thought.

On the Right, though, there was plenty of thought, not new, but ready for a rebrand.

It was the usual story. Deregulate. Everyone will be 'free' to sell their labour at the 'market' price.

That was the factory system during the Industrial Revolution. Welcome back.

*

As I write this, in 2021, the coronavirus has shut down the world economy. We have never seen anything like this before. The only way of saving Western economies has been socialism on the dream-scale. The state is paying our wages, backing business loans, guaranteeing jobs.

At the opposite end of the dream-scale, tech companies have enriched themselves. Amazon is making around $10,000 a second.

Amazon money is made out of a percentage of everything sold on the site. You could say that's what all storekeepers do – and they do – but the store is also a physical and local part of the community where it sits. And the storekeeper pays individual and corporate taxes that in turn help to educate the workforce, keep the roads in good order, contribute towards hospitals, etc.

Amazon's extraction model pays low wages to its workers, absurdly low rates of corporation tax, and, by working globally, refuses responsibility to any locale except its shareholders.

And it's not just Amazon. Google and Facebook employ only a few of us – only 1% of Facebook revenue is spent on wages – but they make money from all of us.

This isn't wealth creation; it is wealth extraction. Just as it is when Uber takes a slice of the money for a cab ride, or when Airbnb gets you to monetise your own bed.

Maybe you make some money by renting out your bed. But maybe your friend is fired from the hotel because it loses business. Or wages are kept low because hotel-room rates must be kept low, because of competition from Airbnb.

At the same time it gets harder to rent an affordable place to live. And those who live close by regularly let Airbnb 'homes'. Do we love it? No, we hate it.

The 'sharing' economy (sharing is not a financial transaction – do words mean nothing?) does not factor social consequences into its business model. The consequences for other people, for thriving cities where people can afford to live, the consequences of endless pointless travel, the toll on the planet, are irrelevant.

The everywhere and nowhere companies of tech certainly deliver to your door – whether it is the information you seek, the friends you are connecting with, the music you want to hear, or yet another package dropped off by a low-paid worker. It looks like a direct relationship between you and what you want, distributed by a benevolent 'service' provider. These companies do indeed provide but the price is high. The lost revenue in local taxes. The local shops closed down. The privacy and anonymity you give away with every purchase and every search and every click and every like.

Marx, finding his way around the Industrial Revolution, urged workers to take control of the means of production.

But what happens when human beings are the means of production?

Or, more correctly, the means of extraction?

Whatever your day job, you, me, all of us are working for the tech companies for no pay. Free stuff is not free. Give your data, give yourself.

Can we regain control of *ourselves*?

Depends on what you believe about human nature.

What we know from the first Industrial Revolution is that people with no education and few resources were able to band together for their own betterment and the betterment of the majority. Unionisation is the power of the collective.

Imagine if the Industrial Revolution had been run as a collective effort between all of the people and all of the planet. No slavery, no child labour, no exploitation, no enclosure, no despoiling the earth. And before you say that could never happen, and the proof is that it didn't happen – yes, I know, but, as I said right at the start of this essay, the past is what we learn from. How we manage the next revolution doesn't have to be a societal nightmare with benefits eventually trickling down from the few to the many.

AI in its multiple manifestations – including automation and robotics – including smart homes and ambient computing – is the technology humans need for the next stage of our development. There's no need to be afraid of the technology – it's how we use it that matters. The invention of the power loom didn't have to create a hateful factory system and slum cities – it could have freed men and women from long hours of work. Instead, working hours increased.

Getting rid of what the great economist and anthropologist David Graeber called 'bullshit jobs' is not anything to mourn. What we need is economic fairness. What we need is to get away from the false binary of sustainability or growth. What we need in the Information Age really is information; not propaganda, fake news, outright lies.

Our problem is that governments do not know how to legislate Big Tech. It is hard enough to get Google, Facebook, and Amazon to pay fair taxes in the jurisdictions where they make mountains of money.

I do not hear of these companies offering to pay a pandemic tax out of their heaped-up Covid profits. Amazon's stock price has risen 70% during the pandemic. Some of its workers have received about 7% extra – but that is deemed a 'hazard bonus' and won't translate into long-term wage increases.

In 2021, Uber was given a High Court ruling in the UK to treat its workers as employees. Uber has fought against this outcome all over the world – and won in the USA.

It's a great idea to connect drivers to passengers – in theory it should encourage the move away from individual car ownership. Uber technology can be on the right side of the planet and the people. But Uber is not volunteering to be the good guy – it costs too much – so legislation is the only way forward.

That's one example. The more I read, the more I realise that the tech companies are, at present, outmanoeuvring their social and fiscal responsibilities to the billions of people who are micro-dosing them with money 24/7.

Accountability matters. In the Data Age, accountability is responsibility. That's what Big Tech has to recognise.

The usual definition of Big Tech is a reference to the top 5 tech companies: Amazon, Apple, Google, Facebook, Microsoft. In reality, Big Tech is about global reach, global control, and a business model that seeks global power without *local* responsibility. Uber and Airbnb are just the same. If that's the model business favours, then the only option is legislation. This won't strangle innovation, as the tech-bullies claim. (Bullies love to play the victim.) Legislation will put a choke on innovation strangling us.

An example: Facebook wants to partner with Ray-Ban to produce facial-recognition eyewear.

Wear these glasses, see a face, and the face-owner's details flash up on your phone – details scraped from social-media accounts. This is already possible via a database app called Clearview AI, which matches faces to personal details.

Did you imagine you owned your face? Owning is so last century. This is a sharing economy. We share. Big Tech collects.

As Tom Paine put it back in 1791:

> A body of men holding themselves accountable to nobody ought not to be trusted by anybody.

From Sci-Fi To Wi-Fi To My-Wi

> These things will make possible a world in which we can be in instant contact wherever we may be. Where we can contact our friends anywhere on earth, even if we don't know their actual physical location. It will be possible in that age, possibly 50 years from now, for a man to conduct his business from Tahiti or Bali just as well as he could from London.
>
> Arthur C. Clarke, *Horizon*, 1964.

The 'things' Arthur C. Clarke is talking about are satellites and transistors.

Let's start with transistors.

In 1965 my dad brought home a transistor radio from the television factory where he worked.

At home, we still had the stately valve-amp radiogram that took up half the parlour, where my mother had listened to Churchill's radio broadcasts while my dad was fighting in the war. As a small child in the 1960s, I liked to sit behind the humming radiogram, watching the orange glow of the glass valves. It was fairy-like and warm.

*

Those valves, as the Brits called them, were vacuum tubes –
the vacuum refers to the airtight space within the tube. The
tubes vary in size from modest to gigantic, and they are shaped
like a baby's milk bottle with a teat. When the cathode (filament) is heated, via an electric current, electrons rush upwards
from the cathode to the anode (metal plate), which is not
heated, but is positively charged. The vacuum ensures that the
liberated electrons can only travel in one direction – towards
the anode. As the electrons are liberated they produce energy
(an electrical field).

The vacuum tube was invented in Britain, in 1904, by John
Ambrose Fleming— really as a spin-off from the incandescent
light bulb, a filament inside an evacuated glass container. When
hot, the filament releases electrons into the vacuum; it's called
the Edison Effect (technical term, thermionic emission).
Thomas Edison had invented the lightbulb in 1879, and
Ambrose realised that if he put a second electrode into a similar
evacuated envelope, like a light bulb, this second electrode (the
anode) would attract the electrons released from the heated
cathode filament, and create a current.

Vacuum tubes become easy to imagine if you think about
old-fashioned filament light bulbs.

Note the teat in the Victorian model – just like a vacuum tube.

Remember (no, probably not, but I am old) how light bulbs used to get really hot? That was wasted energy generated as heat, not light, hence the term, 'more heat than light', and the wonderful expression reminiscent of my entire childhood – 'incandescent with rage'.

Gentle low-energy bulbs just don't offer the same opportunity for 3rd-degree burns or social commentary.

But back to the vacuum tube.

The vacuum tube was the early enabler of broadcast signals, whether the telephone network, radio or TV, and of course, early computers.

Vacuum tubes do their job, but the glass is easy to break, and they are bulky and energy inefficient, as the whole tube gets

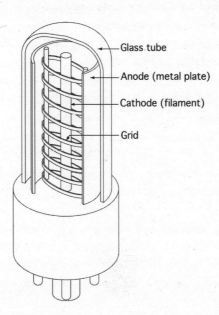

- Glass tube
- Anode (metal plate)
- Cathode (filament)
- Grid

heated up when the cathode is heated up. Early computers were huge because vacuum tubes and miles of connecting wires take up a lot of space, as well as using masses of electricity. The pretty orange glow they give off is waste.

In 1947, at Bell Labs in New Jersey, it was observed that when two separate point contacts, made of gold, were applied to a carbon crystal made of germanium (atomic number 32), a signal was produced where the output was greater than the input. Energy was not being wasted as heat loss. The guys described the discovery as transconductance within the varistor family (varistors are electronic components with a varying resistance depending on the input).

This was a great description for jubilant electrical engineers, but it was never going to sell anything. An internal competition at Bell Labs championed the suffix *ISTOR* as being sci-fi-like and futuristic, and *TRANS* was clear and simple, so the brand-new world-changing product soon became known as the transistor.

These point-contact transistors were developed into junction transistors that can either amplify or direct an electrical signal. For an analogue radio, the signal received from the atmosphere is weak – we couldn't hear it if it wasn't amplified – and transistors were able to amplify this via the built-in speaker.

By the mid-1950s, in America, Chrysler was offering an in-car all-transistor radio – which was better than your wife sitting on the passenger seat underneath a 20lb set of glowing valves.

But it was Sony, in 1957, who manufactured the world's first mass-production transistor radio, the TR-63.

These came in funky colours like green, yellow and red. They looked modern. (Radiograms were brown or cream and looked like your parents' wardrobe.) Best of all, the Sony could fit in a pocket – well, depending on the size of your pocket. The story

went that Sony reps had special shirts with an oversize breast pocket. But, whatever the outfit, the device was cool and neat and contemporary. No cathode meant no glow and no heat-up time. No longer would it take a few minutes, after the familiar click of the Bakelite switch, for the BBC World Service to crackle out of our set.

The TR-63 ran on a 9-volt battery and boasted 6 transistors. Take off the back and here's the circuit board looking like a badly packed 1950s suitcase.

This, though, is the beginning of the future – with the buzzwords we all know and love: *instant, portable, personal.*

By the early 1960s, transistors were replacing vacuum tubes at the cutting edge of technological development. Best of all they were small – and their property of smallness changed everything.

The first transistor measured around a half-inch. They were placed on a printed circuit board. It wasn't until the 1970s that the integrated circuit was developed by Intel – by etching transistors, not onto germanium, but silicon. And then they got smaller and smaller and smaller, like something out of the genie world. So small that your iPhone 12 has 11.8 billion of them.

I think that needs a pause.

6 transistors on the 1957 Sony portable TR-63. 11.8 billion in your hand right now.

But in between then and now, quite a bit has happened – including the moon landing.

In 1969 Apollo 11 landed on the moon. Michio Kaku, theoretical physicist and author, put it like this: 'Today, your cell phone has more computer power than all of NASA back in 1969, when it placed two astronauts on the moon.'

That doesn't mean your phone can fly you to the moon – but it is a useful comparison when thinking about the exponential increase in computing capacity in such a small amount of time.

So, what are we doing with the 100,000 times faster processing speeds in our iPhones?

Well, mainly, playing games. We're smart but we're still apes. Pass the banana.

Thinking of bananas, remember the banana-shaped phone in *The Matrix* movies? The movies that make it seem inevitable that our world is only a simulation? That banana was a Nokia 8110, once the world-leader mobile phone. But not a smartphone. The 1996 Nokia 9000 Communicator was the first mobile phone with internet connectivity – in a really limited way.

Smartphones – digitally enabled devices that can do more than make a call – came into the world via IBM in 1994 with the Simon Personal Communicator. It was clunky, but alongside calls it could manage emails, and even faxes.

30 years earlier, in 1966, in her novel *Rocannon's World*, sci-fi writer and general genius Ursula K. Le Guin had devised the ansible – really a texting/email device that worked between worlds. One end was fixed, the other end was portable. We would be waiting a while for that to hit Planet Earth.

In 1999 Blackberry released their smartphone with the QWERTY keyboard. Like an ansible, with its keyboard and screen, the Blackberry could do calls, but its main function was email.

We had to get into the 21st century for the Apple iPhone.

In 2007, when Apple was already making mega-money with its iPod, Steve Jobs was persuaded to 'do' a phone that would handle everything the iPod did, plus make calls, send emails and texts, and access the internet. To do that Apple turned the humble phone into what Apple did best – computers. Safari-enabled, the iPhone wasn't really a phone at all – it was a pocket computer.

A year later, in 2008 – the year of the global economic crash – Apple added the App Store – which is the beginning of what we think of as a truly smart phone: a phone that is globally connected, and that can be customised (personalised) by the user.

It was a prescient move – a move driven by hackers and developers, who realised that what a phone is for *isn't for* making calls.

Since the revolution in communication that is Facebook, a phone has become primarily a social-media device. Now we go on Instagram, Snapchat, WhatsApp, Twitter, YouTube, play games, check the BuzzFeed, order food and cabs. Google the internet, ask Siri, click on Spotify or Sonos, and sometimes, maybe, make a call. When is a phone not a phone?

Google's soon-to-be-realised dream of ambient computing – really the internet of things, where all smart devices, from fridges to phones, are connected – includes, at a later date, connecting humans *directly* to its services, and to one another – via a nanochip implant in our brain. This will be the ultimate, and planned, end of staring at your phone – an activity that presently involves 97% of Americans and 37% of the world.

The timeline of the smartphone 2007–20** may be one of the shortest in the history of any world-changing invention.

When Arthur C. Clarke was making his predictions of the future to the BBC in 1964, he saw the exponential impact of the transistor, but he also understood that network communication depends on satellites.

The universe is stacked with natural satellites – the earth is a satellite of the sun and the moon is a satellite of the earth. We're talking about man-made satellites here, and the first man-made satellite in space was Sputnik 1.

It looked like a steel beach ball with feelers. The Soviets launched Sputnik in 1957. NASA hastily followed up in 1958 with their own version, Explorer 1. This was at the height of the Cold War between the Communist Soviet Union and the USA and her allies. Anything the reds could do, the West had to do better.

In fact, Britain's famous Jodrell Bank Telescope received the funding it needed to become operational because it was the largest steerable-dish telescope in the world (now it is the 3rd in line) – and it was specifically enabled to track Sputnik 1 by picking up radio waves.

Getting ahead of the pesky Russians was an American priority. It was NASA's 1959 Explorer 6 that sent back to the sharing, caring, democratic free world our first iconic pictures of Planet Blue from space.

Today, there are thousands of satellites in space – mostly put there by nation states for scientific research, including weather tracking and asteroid hunting. Others are for mutual co-operation, such as telecoms, and the global GPS system that tells you (and others) where you are.

TV and phone signals depend on our satellite network; signals are sent upwards to a satellite, and instantly re-located back down to earth again. This avoids annoying signal-blockers, like mountains, and saves thousands of miles of land-routed cable network.

Elon Musk's SpaceX programme, Starlink, controls more than 25% of all satellites in space, and he is seeking permission to get 12,000 up there by 2025, and eventually 42,000. There are risks to all this, including light pollution and energy guzzling. As with so much of tech, most of us just don't know what is going on, and by the time we find out it will be too late to regulate. Musk is aggressively anti-regulation. And who owns space?

Not Elon Musk. This is another kind of land-grab. Another kind of enclosure. Governments will have to regulate space – if they don't, it's already been stolen.

The 1967 Outer Space Treaty declared space to be a common good of mankind.

By 2015 the Commercial Space Launch Competitiveness Act had a different wording: 'to engage in the commercial exploration and exploitation of space resources'.

New technology. Old business model.

A satellite is crazily simple – as well as being enormously complex. Sputnik 1 really is the size of a beach ball. Like every satellite, Sputnik 1 has antennae and a power source. The antennae send and receive information. The power source can be a battery or solar panels.

On the journey from sci-fi to Wi-Fi – when a vision of the future becomes the phone in your pocket – it is transistors and satellites that join the dots.

We think of the computer as the ultimate invention of the 20th century, yet without transistors and satellites your home computer would still be running on vacuum tubes, taking up the whole of your spare bedroom, and you'd be dialling up via your landline.

Are you old enough to remember scrabbling to connect via the telephone line and hearing the wheecracklebuzzbuzzbass of the slow-motion dial-up modem? Actually, it's not that long ago. This is not a light-bulb moment. I live in the countryside, and even in 2009 I had no broadband. I was trying to conduct a love affair with a cool New Yorker living in London. She was fully connected. I was pretending to be. Most mornings saw me propping my laptop on the bread board and running an extension to the solitary phone socket in the understairs cupboard. I made the mistake of leaving the extension in place and a week later the mice had chewed through everything.

Mice love cable. I had no phone and no internet. Progress wasn't on my side.

But what is Wi-Fi?

What's it's not is wireless fidelity.

Wi-Fi started out as IEEE 802.11b Direct Sequence. It's radio waves. Plain old radio waves with a geeky label. Nobody was going to buy into that except a Dalek. So, in 1999, the brand consulting firm Interbrand made a pun on hi-fi, which really is high-fidelity, and came up with the catchy name and icon we all know so well.

In that same millennial-moment year, when we were partying with Prince like it's 1999, Apple launched the first Wi-Fi-enabled laptop.

That is so recent. So near in time.

Broadband internet was city-wide across the world by 2000. That felt like a true new beginning for a true new century.

And look what happened next.

Google had started out as a small search engine in 1998. The telephone-directory-style headline-only internet searches were boring and slow. Stanford students Sergey Brin and Larry Page thought they could do better – and by 2003 Google had become the default search for Yahoo. Google went public in 2004, the same year that Facebook joined the world – or the world joined Facebook.

Those first 10 years of the new century were incredible: Wikipedia 2001, YouTube 2005, Twitter 2006, Instagram 2010.

Even old-forms, like reading, caught the revolution as the iPad and Kindle kicked off mega-sales of ebook publishing.

Those sales and those devices didn't destroy the book though, any more than the car destroyed the bicycle. A physical book, like an apple or an egg, seems to me to be a perfect form. But a perfect form that is still evolving – like the bicycle.

Not everything in this world is destined to be replaced by something else.

What about humans though? Are we going to be replaced – or at least become less and less relevant – or are we evolving?

In the next decade – 2020 onwards – the internet of things will start the forced evolution and gradual dissolution of Homo sapiens as we know it.

But before we get to the internet of things and a world of connected devices – and some directly connected humans – let's go back to the internet itself – to see how far we have come, and where we might be going.

Back in late-1960s America, soon after the Summer of Love, the Advanced Research Projects Agency Network (ARPANET) adopted a British packet-switching system to transmit limited data between research institutions. At the same time the TCP/IP protocol suite was established.

The more familiar term, INTERNET – really just inter-networking – came into use in the 1970s, to describe a collection of networks linked by a common protocol.

It was Tim Berners-Lee, an Englishman working at the physics lab CERN, in Switzerland, who developed HTML. HTML (hypertext mark-up language) allowed hypertext documents to link into an information system accessible from any node (computer) in the network.

In 1990 the World Wide Web as we know it came into existence. Think of the internet as the hardware and the web as the software. By 2010, the web had become the way for billions of us across the world to access the internet.

And, of course, we had Google as our search engine. The bigger the internet, the more sophisticated the search needs to be. The question now, though, is, are we being nudged as we search? How neutral are our finds? What does the information reinforce? What is excluded? What bias operates?

Do we really want advertising in our face whenever we type in a word?

Do we want our data tracked and traced and repackaged? Do we want to be profiled by an algorithm?

Why can't I buy something online without clicking ACCEPT on their privacy policy – which really means what I have just bought is not private at all?

Personalising the web is where the money is. Your web – where everything is tailored to 'help' you navigate faster, get to what you want, often via what you might be persuaded to want – is the new consumer model where the customer pays twice – with cash for the goods, and with the free gift of information about ourselves. That information is valuable. Even when we aren't buying stuff, when we are browsing around, or using social media, we are being strip-mined for our data.

Data profiling allows accurate and individual targeting. Ads aren't just selling you any old stuff – they are trying to sell you stuff your cookie trail tells them you might be persuaded to buy.

More worryingly than selling you stuff, your newsfeed is algorithmically tailored to what you 'want' to hear about. Our clicks and likes determine the so-called Editors' Picks, making sure that the little we know – and all our personal bias – will be looped back to us again and again, ensuring more clicks and likes in the echo chamber of 'choice'. Access to different ideas and a wider world view just disappears. It's censored. Not by a censor, of course, because that would be totalitarian – but by what looks like personal choice, your very own personal choices, nudged a little, just for you.

Most of life is about being wrong, making mistakes, changing our minds. Web profiling means you need never be wrong, never seem to make a mistake, never have to change your mind. You'll be sold what you have already bought. You will read what you have already read. Amplified. x 100.

*

This will get more interesting/worrying as Siri and Alexa grow up – or if Google figures out how to develop a genuine personal assistant for each of us.

Siri and Alexa are fun, but all they really do is connect with your existing systems – open Amazon Storytime, play music from your Sonos library, reorder your cat food, turn on the Nest thermostat controls, and search the web multiple times faster than you can.

An AI PA will be a mini-me – a neural network (a series of algorithms trained to recognise patterns) that learns my wants and needs, my favourite foods, my travel preferences, restaurants, calls, bills I forget to pay, birthdays I forget to remember, and all my digital photographs, my texts, my emails. My self wherever I have hidden it. And what about my politics? My dirty secrets? Even my thoughts?

Here's Larry Page on how Google will go in the near future:

> Our ultimate ambition is to transform the overall Google experience, making it beautifully simple. Almost automagical, because we know what you want and can deliver it instantly.

And while what you want is being known and delivered, all of that will be tracked. Cookies, remember, are bits of tracking code inserted into your computer. When you, personally, are no longer doing the searching, because your personal PA is doing it for you, there will be no effective privacy preference. But in fact there really isn't one now. Even innocent-seeming apps like the weather and ride-sharing are infested with tracking code.

A mini-me PA will be a seductive choice.

Why wouldn't we want an able, considerate, smart helper who is always available, and mostly free? That used to be called a wife. But then feminism spoiled the party.

She or he could, of course, in time, be a double agent. In a *Blade Runner* world, I could be turned in to the authorities by my own virtual mini-me.

And she'll know where the money is. Where the bodies are. Who my friends are, and how to find them.

Can I run away? In a cashless world I will be using my phone to pay for everything at first – and then iris recognition, or fingerprints, or chip implant will do away with the need for external devices. I will be my own device. No need for a wallet, or a phone, or a set of keys, or an office-swipe. I will be free. And followed everywhere. Or, rather, there will be no need to follow me because my location will be obvious from whatever satellite is my own personal star.

Sci-fi. Usually a dystopia for the hero and friends, and a utopia to those who benefit from or accept the situation. I suspect that the future will not be so binary. As Arthur C. Clarke put it in 1964 – 'The future will not be merely an extension of the present.'

What's for sure is that 'privacy is an anachronism' (thank you, Mark Zuckerberg). Like every other system in the coming world, I too will always be on, always be known, always be available, even while I sleep, dream, or think about things. My 'own' self will not be owned by me. My own self and my known self will become the same data set. Soon the interface between me and mini-me – my Google self and myself – will be redundant. We will merge.

It will be easy to read our thoughts. A smart implant works both ways.

One of my favourite sci-fi novels is John Wyndham's *The Midwich Cuckoos* (1960), made into the 60s movie *Village of the Damned* – and again in 1995, directed by John Carpenter.

Sky has it in development now for an 8-part TV series written by David Farr (*The Night Manager*), so it's pretty clear this story has ongoing impact.

Every woman of childbearing age in the English village of Midwich gets pregnant and gives birth to a blonde child. These children grow up fast and are spookily clever – and they can communicate with one another telepathically; in effect, the children are connected neural networks.

They can mind-read others too. Nothing is hidden from them. They claim their intentions are benign. But who knows?

Think about the 10 clicks on Facebook that will deliver a profile of the user more accurate than a mother or partner can deliver. More accurate than what we think we know about ourselves.

Think about a world where there are no private thoughts.

Think about a world where there are no private actions.

The internet of things will allow any object to act as a computer. Your fridge will tally the food you buy and eat. If you sign up to a dieting app, the fridge will 'help' you by directly ordering the food you should be eating. The fridge will also self-lock if you break the rules. Desperate folks trying to hack their own fridge is a new torture coming your way soon.

Smart beds will be able to monitor – and assist – a good night's sleep, warming or cooling the bed, managing light flow, and reporting your state to your automated doctor; you might need medication. You might not be fit to drive today. Did you have sex? No? Is your relationship healthy? Perhaps you need a counsellor, or Viagra.

In the smart kitchen, the toaster will remind you that today is a no-carbs day. The smart toilet will assess the contents of your initial evacuation. (I am not making this up.)

And what about cars? Driverless cars are a lovely idea – we can go to the pub, act like we've got our own chauffeur, work in the back seat. There's no cab driver giving us his worldview of American politics, just peace and quiet, and our own private pod.

But what happens when you realise that this lovely private pod isn't private at all? Driverless cars track your movements.

Random becomes predictable becomes dossier. The pods can be redirected. Police station? Your scary ex's home? Enforced trip to hospital as your implant registers your blood-pressure risk? Hijacked? Kidnapped?

Autonomous isn't anonymous.

And driverless cars will have to be programmed for tricky situations. Try this game: an unavoidable crash scenario means you are going to run down a little kid or an old lady – which one? And do we favour pedestrians over passengers? The least damage to me in the pod or to you on the pavement? What if a dog runs into the road, and the car needs to brake, at the same time as I need to accelerate to stop someone shunting me from the rear?

In such situations a human makes a split-second decision. Autonomous vehicles will have to be programmed ethically – if that turns out to be the right word – ahead of time, and they will run according to programme … until the spooky day when they rewrite their own programme.

And drive us all off the cliff we deserve.

Cars you can still pilot yourself will be factory-fitted with smart sensors. These will be able to listen in to your conversations, your radio stations, to detect your alcohol or drug consumption. Even your mood. If you start having a fight with your boyfriend, the car can call the cops, or direct you to pull over, with a warning that your insurance rating is in jeopardy. Telematics are very interesting to insurance companies; the crude black boxes fitted to monitor the road behaviour of rookie drivers will be swishly inbuilt in new cars and with a whole range of new features. Speeding? You could get an instant fine. Road rage? All recorded.

Where were you last weekend? Your car 'knows'. Coercive control has never been so simple.

Plus, if you don't pay the instalments on your car, the company can remotely disable the engine, locate the car on GPS, and tow it away.

*

Leaving aside the monetisation of every breath you take, there
will be advantages to being seamlessly connected.

Chore-work and bore-work can be taken care of – who
wants to go to a supermarket, report the faulty boiler, wait in
for the plumber, or manage their health, when check-ups with
the GP can be made by your monitoring implants, and smart
homes will run themselves, including taking receipt of goods,
and letting in the plumber, whose movements will be visible on
your phone and whose access will be strictly timed?

Instead of wasting time on the above, discuss with your
mini-me PA what vacation is best for you right now. Then she
will book it. When you are stressed she'll suggest a night off, or
a night out, and participating restaurants will deliver the meal,
or the car will call by to take you. Don't even think about it. It's
taken care of ...

Companies hope such advantages will be seen to outweigh
the loss of personal sovereignty. We spend a lot of time wasting
time – it may be personal, but it's hardly sovereign.

Anyway, all those self-help books and apps suggest strong
incentives to turn ourselves into someone else. Someone
happier, more efficient, better ... what the internet of things
promises to deliver. And in many cases will deliver.

When you are gaining so much, does it matter if there are no
secrets, and perhaps no self, anymore?

There is a whole philosophical mindset within tech companies, and supported by their gurus, that believes Enlightenment
ideas of the individual such as free will, autonomy of self,
self-direction, genuine choices are bunkum. Humans are not
solitary, self-directed creatures. We need to be known. We are
social animals. We like groups. We like belonging. And, as
such, we are readily influenced. Our behaviours are learned and
can be modified. We mistake habit for choice. We believe
received opinions are actually our own. We are also lazy, and

prefer an easy life – we like being 'told' what to think, as long as we can believe we are thinking for ourselves. Militant Trump supporters made this frighteningly clear as they stormed the Capitol in the name of freedom, obeying orders from President Trump.

QAnon supporters have no basis for their extreme views, and yet they believe that they are thinking for themselves – rather than being manipulated.

Anti-individual behavioural theory came into vogue with the American Harvard psychologist B. F. Skinner after World War Two. Back then it was straightforward Behaviorism. Now it's Radical Behaviorism, and there are plenty of psychologists helping companies and political groups with their moral reasoning here.

Steve Bannon in the USA and Dominic Cummings in the UK are disciples of the new-style Skinnerism that seeks to manipulate 'individual' behaviour. (Possible because such behaviour isn't individual at all – it's an amalgam of upbringing, bias, assumptions, and fear, reward being the flip-side of fear.)

Nudging our behaviour towards desired outcomes, whether it's what we buy or how we vote, has always been the standard fare of lobbying and advertising. Social media has vastly extended the reach of the exercise. The behavioural data that companies and lobbyists need is handed over, by its owners, for free, every second.

Facebook seems to have proved that 'sharing' your whole life with others is not just the price of connectivity, but its purpose and pleasure. Instagram and Tripadvisor usage suggest that it is the *sharing* of experience, and not the experience itself, that is most important. The experience once shared, a whole host of people with no previous desire to visit an obscure Cotswold village, or eat at a particular family restaurant, suddenly Insta and tweet and Facebook their version of the moment – which is then repeated by the next wave of people who never wanted to do this – ad nauseam, into a nightmare cartoon of sheep-like, herd-like, borrowed reality.

Skinner's end-of-life observation in 1990 that 'a person is simply a place where something happens' seems to be confirmed by social media.

That a person might not be a place at all, but a carrier of history, a second chance at the future, a being capable of love, a moment that is not capturable, an interior force, a private act with public consequences, but not ultimately public – in the way a park or a shopping mall is public – is what? Romantic? Foolish? Wrong? Or a view of the self that is worth sustaining?

In her excellent book *The Age of Surveillance Capitalism* (2019), Shoshana Zuboff, professor emerita at Harvard Business School, makes the case for a different kind of future human – one that is participatory, not behaviourally modified. A future where technology is a tool for the greater good of all, not only for the money-makers and decision-takers. A future where democracy still has a place, and where Facebook, Google, Apple and Amazon are not carving the world up between them just as the Imperial powers once did.

The new reality, though, will not be sold to us as surveillance, with its totalitarian overtones. The future will be sold to us as empowerment.

Elon Musk's Neuralink company is working on brain-computer interfaces – threads that will allow someone to control a computer via their thoughts.

Human trials started in 2020. The current aim of the tech is to help people with paralysis, a laudable aim. Musk's eventual aim, though, seems to be symbiosis with AI, so that humans don't get left behind in the intelligence game.

*

Steve Austin: the original bionic man in the 1970s TV series. The faster, stronger, smarter guy whose biological limits had been reset.

Modern medicine has already reset human biology. We live twice as long as our ancestors at the start of the Industrial Revolution. In 1800, life expectancy was 40 – and the early factory system, with its hellhole slums and 12-hour days, cut that by around 10 years.

200 years later, 80 years has become the new normal (though even in the Bible we get offered threescoreyearsandten, so it's taken a while to pass home-base, whatever the enthusiasts for Capitalism have to say).

Those of us who have a good diet, who exercise, who have access to healthcare, and not too much stress, can live healthily as well as long. The rich, who have access to the best of everything, are doing very well. Naturally, they want to do better, which is why Silicon Valley is investing in research that will stall, or reverse, physical and cognitive decline. Cognitive decline for humans is as real as muscle loss and organ failure. In the end, our biology beats us.

AI systems, embodied or not, suffer no such losses and no such decline. AI systems can augment, version-up, get smarter. If humans are, say, HomoSapiens3, the post-Industrial Revolution version, we shall have to get to HS4 pretty soon if we want to stay in the game. Merging with the AI we are developing is a logical outcome. Outcomes, though, often resist prediction.

What can be predicted is *personalisation*.

As we saw, personalisation began with that first transistor radio back in the 1960s: at last, a small portable device just for you – no need to sit round the family radio. Go your own way. Personalisation was enthusiastically adopted by the laptop and smartphone industry.

Simultaneous with our fully public-data-harvested-known selves is the personalisation of that self. It will be 'your' smart implant. 'Your' smart car/house/lifestyle/insurance/portfolio/ personal shopper/fitness guru/therapist/PA. Tailored to you, your tracker-helpers will change and develop seamlessly (frictionlessly) as you do.

That's a clever marketing move. Personalisation isn't just about the product anymore. It's the concept.

Personalisation is being offered in place of the old-fashioned outmoded idea of privacy.

But why is privacy problematic in our internet-of-everything future?

Privacy is friction. In economics-speak, friction is the opposite of flow. Friction is whatever impedes the data flowing from you and all that you do, to the interested parties who want to make money out of you, and/or control/nudge your behaviour. It is as simple as that.

Or am I just an analogue human who likes the idea of being off-grid sometimes? At present that is still possible if you leave your phone at home, walk wherever you want to go, pay cash where there is no CCTV (both increasingly difficult moves, I admit), don't browse the internet for a few days.

Soon, though, as smart devices and smart implants become normal, you won't be logging in. You're in. You're on. For life.

From sci-fi to Wi-Fi to my-wi.

Former CEO of Google Eric Schmidt, sitting next to Facebook COO Sheryl Sandberg at Davos in 2015, put it like this:

> The internet will disappear. There will be so many IP addresses, there will be so many devices, sensors, things you are wearing, things you are interacting with, that you won't even sense it. It will be part of your presence all the time.

This raises new ethical questions.

Neuroethicist Marcello Ienca (ETH Zürich) has proposed four rights for the technology age:

1)   The right to cognitive liberty
2)   The right to mental privacy
3)   The right to mental integrity (protection from brain-jacking)
4)   The right to psychological continuity

Number 4 is especially interesting to me because it suggests that the removal, as well as the implanting, of neurotechnology will be an issue, perhaps a threat, or even a new torture (we can downgrade you/make you crazy/rewire you etc).

If/when we get to the point where uploading our whole brain to a computer is possible – so that we are 'stored' – who is to say that on the way to storage, certain memories/experiences won't be removed or others added in, Philip K. Dick-style in his brilliant short story 'We Can Remember It For You Wholesale' (the movie *Total Recall*)?

For young people – digital natives who have grown up with a phone and Facebook, whose every move on Insta, and who want to be influencers themselves – questions about protection, privacy, content-screening and platform responsibility are being asked by an NGO called 5Rights.

5Rights was founded in the UK in 2015 by filmmaker Beeban Kidron. When I asked her about 5Rights, she pointed out that over a billion underage young people are online every day, for hours every day, and treated by the platforms they use as if they are adults. Content is easy to access, hard to monitor. Online grooming is a particular threat.

For example, during the 2020 Covid lockdown, in the UK (just the UK here, folks) around 9 million attempts to view child-abuse images were blocked in one month alone.

Kids are tech savvy but tech vulnerable. 5Rights wants to see kids protected in the digital world just as they are in the physical world. In the physical world we do make a distinction between children and adults – a distinction hard-won during the Industrial Revolution. We don't want our kids working in sweatshops, but we seem unconcerned about exploitation via their phones.

That includes addictive gaming and porn habits, as well as the suicidal misery of 'likes'.

Data collection that starts early in life amounts to a conquest of that life. And, as we have seen in the stand-off between China and Hong Kong, forced data removal from popular sharing sites like TikTok can be used to persecute or prosecute young people, to monitor their behaviour, and no doubt to influence their political 'choices' later. In China's case the data-snatch is clearly political. That's not the point though. China is doing in an obvious way what is being done quietly and covertly in the 'free' Western world every day. Our data is not anonymous. Who has a 'right' to this data? To sell it? To snatch it? To package it?

5Rights has a good chance of worldwide co-operation and legislative success as far as the big platforms are concerned, at least in the way they are operating at present.

What happens, though, when there is less, or even no, division between the online worlds and the physical world? When Eric Schmidt's prediction of the end of the internet and the start of the internet of everything happens? How do we protect anyone when the internet is always on and we are always on it?

I loved that great story of the online-addict teenager whose mother confiscated all her devices and turned off the Wi-Fi. The kid realised she could send tweets from the home's smart fridge. The whole thing may well have been a hoax, but Reddit offers full instructions on how to tweet from a Samsung fridge, if you have one.

The point of this story is that the goal of our digital masters in Silicon Valley is that there will be no offline. No need to hack your fridge.

And yet …

It may be that all of this – privacy, protections, dataflow, and data usage – is a temporary problem. At present we imagine human interests and human actors as dominant in all our scenarios. If, though, AI does become superintelligent – a player and not just a tool – then the future for humans may be irrelevant. I mean, how much data will AI need on a species being consigned to the Museum of History?

When I talk to people about the future, many believe that the world's head-in-sand attitude to climate catastrophe will make that other kind of sand – silicon – irrelevant, in or out of a valley. We'll be fighting for food, not tweeting from our fridges.

Others believe that developing super-intelligent AI, as swiftly as we can, is our best chance at survival.

Until 2020, none of us was thinking about viruses as the wipe-out call. Now we are.

Ironically, although the world may become much poorer because of Covid-19, the virus is a chance for the tech giants to get much richer, and to get more control. And not just Amazon doing home delivery.

Eric Schmidt has been talking enthusiastically about home-schooling for all (except the rich, you can be sure), using the platforms that have begun to replace contact during the pandemic.

If we are at home, we will need to be connected in new ways. That's an opportunity for the connectors. Virtual-reality avatar sessions are being trialled by Facebook. I am sure they will become as real for us as Zoom is now.

More worryingly, track and trace everywhere, including even a visit to the pub, is going to allow levels of surveillance that civil-liberties groups would have taken years over, arguing about privacy and usage. That's all gone now. Being watched equals being safe.

But what about energy constraints? AI is an energy guzzler, and even if we extracted all the fossil fuel left on the planet it wouldn't be enough for the kind of super-future envisioned by Ray Kurzweil or Elon Musk. That's why the premise of *The Matrix* is that humans are just battery packs in an AI simulation.

Optimists say that the energy requirements of an AI future will force the world towards low-carbon solutions. The market will drive the change because it must.

But there are other constraints on an AI future too.

Intel co-founder Gordon Moore came up with his own law: Moore's law observes that every two years the number of transistors that can fit onto a square inch of microchip will double. 50 years after that first Intel chip, the computing power that once filled a whole building with hardware now fits into your handbag. And it uses much less power. That's progress.

There is a limit to progress though – and unless we switch our systems we are pretty much at the limit. Put simply, there's no more room, at the small scale of the laptop or phone, to keep doubling the number of transistors. No matter how tiny, they still take up (some) physical space. The next jump to faster processing speeds and more commands will be quantum computing.

There's a rumour that China has made the breakthrough already. If they have, they aren't telling anybody. Google and IBM are both claiming to be nano-close.

Transistors work on the familiar zero-and-one principle – whether it's analogue or digital. Zero and one is how classical computing performs its calculations. 'Bits' of information each hold a 1 or a 0. A quantum 'bit', or qubit, is different. Very different. By harnessing subatomic weirdness, a qubit can be a 0 and a 1 *at the same time*. That's because in the world of the very small (or the very cold), states are not defined until they are measured. They exist *simultaneously* in separate and indeed contradictory states. Only when observed (measured) do they take on a definite form. This is fine in magic, and every magician and fairy tale employs the trope of this simultaneous power – and maybe that's why we do deeply understand that the reality we live in, defined and measurable, is only one reality, and a bit clumsy, and on the surface.

Anyway, in the world of subatomic weirdness, binary isn't real or relevant, and this leads to astounding possibilities of power.

Numbers-wise, 8 bits make a byte. Your smartphone memory could have two gigabytes – that's 2 x 8 billion bits – but a few dozen qubits is way, way beyond that.

According to Dario Gil, director of IBM's research unit in Yorktown Heights NY,

> Imagine you had 100 perfect qubits. You would need to devote every atom of Planet Earth to store bits to describe the state of that quantum computer. By the time you had 280 perfect qubits you would need every atom in the universe to store all the ones and zeros.

At present, IBM Q System One lives like a reclusive rock star inside a 9-foot cube of black glass only accessible through 700lb doors a half-inch thick. Quantum computers must be absolutely remote from any entanglement with reality – entanglement affects the outcome, as anyone who has ever fallen in love knows.

So, we are building a god so remote that it must live in an inaccessible temple visited only by high priests in special clothes. The high priests can ask the questions and interpret the answers.

Quantum computing may be the future, but that story is like a Pharaoh-dream from the past.

Where is all this heading?

What's for sure is that fewer and fewer people will know how the systems that control us actually work. We're not talking here about fixing the washing machine.

While there is no consensus on the future, there is consensus that total connectivity will happen – to the internet, to our devices, to our machines, to each other.

When you customise your connectivity it will feel like yours. Actually, it will feel like you. And you will feel as if you have chosen it. An avatar Sinatra singing I did it My-Wi.

And a whole new questioning of what is 'you' comes bubbling up. My-wi is religious in its own way. Mark Zuckerberg has talked about Facebook as a 'global church', connecting people to something bigger than themselves. It may be bigger – it may be smaller – but it will be connected.

Here's George Orwell in his novel *1984*:

> You had to live – did live, from habit that became instinct – in the assumption that every sound you made was overheard, and, except in darkness, every movement scrutinised.

Orwell might have been surprised to see *Big Brother* become a TV show – and with infrared cameras; not even darkness is a cloak of invisibility anymore.

He might have been more surprised to watch people competing to go on surveillance shows like *Love Island*.

*

The success story of Homo sapiens has been one of infinite adaptability. Adapting ourselves to the Machine Age was an unprecedented rip with our evolutionary past. We mourn the price to Planet Earth, but few would wish to return to a pre-1800 world. We dislike the Intrusion of Everything, but who would want a world without smartphones and Google?

Perhaps, though, we would prefer a world that may be less democratic, but that could also be less stressful for the next stage of our development.

My-wi might leave us as little children are: cared for, fed, safe, watched over, with plenty of fun stuff and free stuff, and with someone else deciding the big stuff.

No reason to believe that who decides will always be in human form.

AUTHOR'S P. S.

On 22 May 2021 I read in the *Financial Times* that the next big thing for blockchain and cryptocurrencies will be DeFi – standing for decentralised finance – how to bypass intermediaries in financial services transactions. You might want to do this if you are breeding or selling digitally created racehorses (NFTs) . . . Recently one such horse sold for $125,000. These horses don't need feeding – but you can race them, bet on them, and yes, breed them using algorithms.

Zone Two

What's Your Superpower?

How Vampires, Angels, and Energy Reimagined Matter.

Gnostic Know-How

If you understand something in only one way, then you really don't understand it at all. The secret of what anything means to us depends on how we've connected it to all the other things we know.

Marvin Minsky, *The Society of Mind*, 1986

Marvin Minsky (1927–2016) was a mathematician who co-founded the AI lab at MIT (Massachusetts Institute of Technology) in 1959. His work on robotics and neural networks led him more deeply into research about what intelligence actually is – and the possibility of independently intelligent machines.

Could AI be more than a tool invented and used by humans?

Minsky was one of the advisors on Stanley Kubrick's 1968 movie *2001: A Space Odyssey*. Remember HAL, the creepy computer who takes over the mission with an agenda of its own?

Computers were still filling up room-sized spaces back then, but by the late 1950s practical application of the physics of semi-conductors meant that transistors were on the way to replacing the vacuum tubes that made computers so bulky.

There was plenty of excitement in the 1960s about the shrinking size and increasing power of computers – these breakthroughs made possible by transistor technology – but aside from the practicals, that feeling of excitement was the sense that what humans were developing with computers was unlike anything else humans had developed before.

Not just another brilliant invention. Something different to planes and automobiles. Different to TV and telephone.

Humankind's latest brilliant invention might turn out to be our 'last invention'.

That 1965 phrase was coined by Jack Good, one of Minsky's associates on the Kubrick movie.

Jack Good (1916–2009) worked with Alan Turing at Bletchley Park during the war, helping to build the computing apparatus that would eventually crack the codes used by the German Enigma machine.

In the 1980s, Jack Good and Marvin Minsky demonstrated that unsupervised artificial neural networks (mini-brains) could learn by themselves, and self-replicate, separate to human input. So, it was only a matter of time – *when*, not *if* – before machines could manage themselves, without humans doing the programming. The problems Minsky and his associates encountered were problems with computing power and speed – not with the theory of autonomous neural networks.

Jack Good put it like this:

> Let an ultra-intelligent machine be defined as a machine that can far surpass all the intellectual activities of any man however clever. Since the design of machines is one of those intellectual activities, an ultra-intelligent machine could design even better machines; there would then, unquestionably, be an intelligence explosion, and the intelligence of man would be left far behind. Thus, an ultra-intelligent machine is the *last invention* that man ever need make, provided that the machine is docile enough to tell us how to keep it under control.

It was that last sentence that prefigured the not-docile, not-under-control HAL 9000.

Marvin Minsky credited Alan Turing as the mind that prompted mathematicians to consider a computer (inside or outside a robot) as more than a programmable tool.

Turing, with his famous Turing Test, believed that a machine would, one day, prove indistinguishable from human intelligence. He believed that day would come sooner than it has done – he suggested the year 2000 – but he was sure that a machine could display human-level interpersonal skills.

Minsky credited his own enthusiasm for developing an AI system that can think for itself to Turing's 1950 paper *Computing Machinery and Intelligence*.

Read it now, and to me the most interesting part of a fascinating paper that includes telepathy and hive-mind networking is what Turing called *Lady Lovelace's Objection* – replying directly to the dead genius who 100 years earlier had concluded that Babbage's Analytical Engine, while being theoretically able to write novels and compose music (a big insight in 1843!) would not be capable of *originating* anything.

This needs a bit of clarification.

Ada, following her father, the poet Lord Byron, believed that *originating* meant an entirely new insight or form not traceable to pre-existing material, unlike the idea of feeding in vast quantities of data and seeing what comes out the other side. We can all argue this definition, of course, because humans themselves are data-processing plants, just as computers are. But you get the point Ada was making.

Turing got the point. Most of his colleagues agreed with Ada, without acknowledging it was Ada they were agreeing with, because Turing was the first scientist since her death to take her seriously in her own right – rather than as a footnote to Babbage. He read what she had to say, and he asked himself:

Is Lady Lovelace correct?

Can machine intelligence be said to originate?

How does machine intelligence differ from human intelligence?

At the end of World War Two, Turing went to work in the fledgling computer department of Manchester's Victoria University – what is now the University of Manchester, UK.

There, with Max Newman and Tom Kilburn, Turing worked by day on the practical problems of building a stored-programme computer. At night he dreamed of beautiful men and beautiful machines. Machines that could do more than crunch numbers or play chess. Machines that could talk to you. Think with you. And one day, perhaps, think beyond you. Turing was ahead of his time, just as Ada Lovelace was ahead of hers.

Arrested for gross indecency in 1952, Turing didn't live to see, or to work on, the astonishing breakthroughs of the coming decades. He was dead – almost certainly a suicide – in 1954. What Turing wanted to do with his body – have sex with men – seemed to be more important to petty, post-war Britain than what he could do with his mind.

It was the nightmare of Oscar Wilde all over again and 50 years later. Humans, unlike machines, learn slowly, and, unlike machines and bees, fail to apply what they learn across the 'hive'.

Turing's friend Jack Good said, 'I'm not saying that Alan Turing won the war, though I am sure that without him, we might have lost it.'

It is impossible to overestimate the impact of the Bletchley Park team on the future of computing. As the war ended in 1945 the new future, the modern future, our future, was beginning.

Looking through a list of significant world events in 1945, it's dominated by the end of the war, the founding of the United Nations, Britain's first Labour government, de Gaulle in France, Gandhi's fight for Indian independence, plans for the future state of Israel, America's Marshall Plan to reboot the world economy. The big politics that must follow a world war.

In 1945 something else happened too. An odd discovery that seemed, at the time, only to have a bearing on the foundations and formations of the Christian church – and surely that was long enough ago not to matter in a new world trying to begin again after a world war. But if we take Minsky's view that knowing things in more than one way is what is important to systems of human knowledge, then I think the following story is worth telling.

North-west of Luxor, in Egypt, is the town of Nag Hammadi.

In 1945 two peasant farmers were out in a cart digging up mineral soil for fertiliser. One swung his mattock and hit what turned out to be a sealed jar nearly 2 metres tall.

At first the brothers were afraid to open it in case a genie lived inside.

But what if it were full of gold?

Curiosity overcame fear. They smashed the jar.

Inside were twelve leather-bound papyrus codices, written in Coptic, probably translated from Aramaic or Greek originals, and dating from the 3rd and 4th centuries, though one of them, the Gospel of Thomas, may be dated as early as 80 years after the death of Jesus Christ.

The books were mainly Gnostic texts.

Gnosis, in Greek, means 'knowledge', but not the knowledge that comes from study; rather, a knowledge of the ultimate essence of Self and World. We know the word in English from agnostic, and Albert Einstein called himself agnostic – someone who doesn't 'know' rather than someone who doesn't believe in the existence of God, in whatever shape or form.

Gnosis isn't science; science depends on objective measurement and repeatable demonstrations, and how do you measure a deep sense of what is known – that can't, at least at the time, be proved by any available method or metric?

Gnosis isn't religion either, in that the Gnostics neither wanted nor needed creed, doctrine or hierarchy. The Gnostics weren't interested in building a church with a clear set of beliefs; they wanted to explore *the nature of reality*. That put them at odds with the increasing power of institutional Christianity – so much so that around 180 CE Bishop Irenaeus declared Gnosticism a heresy and issued a decree that any Gnostic texts be burned. That is probably why, and probably when, the Nag Hammadi jar was buried.

So, what did the Gnostics believe, and why is it relevant to the coming world of artificial intelligence?

For a Gnostic, the world is not a once-upon-a-time perfect place in need of redemption. There is no Golden Age. No Good Old Days. No Paradise. No Fall. Our world was created badly from the start – not out of evil, but out of ignorance.

The story goes that somewhere in the universe is the Pleroma, the place (but it isn't a place – it's a concept) of fullness and light.

In the Pleroma live the aeons (the timeless ones). The aeons do have a commander-in-chief, a kind of godhead, but otherwise they work in pairs. This part of the story takes its influence from the Greeks, who believed in the godhead as a dyad, not a monad. So 'male' and 'female' are unified as well as separate.

In fact, the aeons work like the zeros and ones of code. What is really interesting is that they work like quantum bits (qubits), because they are simultaneously both 1 and 0 – but that's another story ... (Recognise it?)

The least experienced, most junior, of the aeons is Sophia. Her name means 'wisdom', and she appears in the Old Testament too, notably in the Song of Solomon. Obviously, she is a goddess-figure who in the strict monotheistic patriarchy of Judaism got lost or demoted. (In the Trinity, the Holy Spirit is really Sophia).

Sophia broke with her aeon pair and set out to create something all by herself. What she created was a steaming pile of matter, and a swaggering, bullying demiurge called Yaldabaoth. It was all a terrible mess. Yaldabaoth plunged his clumsy hands into the pile of poop, and formed a world out of this sweating, semi-alive playdough. Then he made some people. Then he decided he was God. In some texts he is called Jehovah.

It's a Trump-style story from the beginning of time.

As part of a pair, Sophia isn't alone with her mess. Her counterpart, Christos, comes to pull her out of the bog of dark matter she has created. When they discover that this steaming, sweaty, material world is still infused with a glimmer of divine light, Christos agrees to impart the necessary knowledge (gnosis) whereby humankind can be informed, *Matrix*-like, of their true origins and true home.

Humans, therefore, are not sinful. Humans are ignorant.

Discovering who we are and where we belong is our task.

All of those fairy tales where the princess is locked away and needs to be rescued come from this source – as do those stories of going to seek your other half, like Orpheus and Eurydice, or the mission of righting a wrong, like St George and the Dragon. It has all got tangled up with boring binaries of male and female – active and passive – but once you remember that Sophia isn't part of a binary, but is an equal and active qubit, it makes sense.

Gnostics agreed that being made of meat is ridiculous.

For a Gnostic, the mind/body split that came to dominate Western thought after Descartes, in the 17th century, was understood as trapped light. Not soul and body. Our non-biological spark wasn't a soul in need of redemption, in the Christian sense – it was a mind in need of knowledge.

Ignorance, for Gnostics, was a form of self-destruction. A narcotic. An addiction. They understood the resistance to gnosis as the easy wish to stay asleep – or to be drunk all day – or play games. Gnosis is hard work. Gnosis is upsetting. Think of it as the choice between the blue pill and the red pill in *The Matrix* – essentially a Gnostic movie.

The end of ignorance is not to be found in creeds or rituals or doctrines of salvation – or from instruction by priests; it is to be found within. There is a lovely piece in the Gnostic Gospel of Thomas:

> Knock on yourself as upon a door and walk upon yourself as on a straight road. For if you walk on the road it is impossible for you to go astray. Whatever you will open for yourself, you will open.

Gnostics were friendly towards women and suspicious of hierarchy. Unlike either Judaism or orthodox Christianity, most Gnostics treated women as equals and did not seek power structures. Both of those positions drove Bishop Irenaeus mad – and were evidence to the orthodox of the madness of the Gnostics (in Greek, *ortha* means 'right' or 'correct'; *doxa* means 'the way').

The orthodox view held that women were subordinate, and that hierarchical power structures were necessary for discipline, order, continuity and succession.

Yet Jesus himself had broken with Jewish religiosity in his treatment of women – he spoke with them and ate with them as he did with men, and Mary Magdalene, the whore with a heart, was always near him. Nor did Jesus aim at power. He was a homeless man dependent on the generosity of others. He had neither possessions nor money.

For the Gnostics, Jesus the man was not the Son of God, except in that we are all children of God. Some of the texts found at Nag Hammadi depict Jesus as a non-human (Christos). A being made of light. Not meat.

Of course, frustration with embodiment led some Gnostics to hate the body – but plenty of Christians and Jews hated the body too, as a weak and corrupt vessel. Hatred of the body has increased over time. Probably now in the West, in the 21st century, we hate our bodies more than at any other time in history. We diet them, abuse them, surgically transform them, spend billions altering whatever we can – and billions more trying to live as long as we can. Modern medicine treats the body as a site of perpetual intervention – a thing that can't be trusted to manage itself.

And yet the human body is an everyday miracle. We shall miss it when it's gone.

Maybe …

For the Gnostics, death did not lead to salvation or the lack of it. There was no final judgement, or final destination, as in Christianity and Islam. Either light returned to light – following the Aristotelian view of a world-soul to which we all return – or, following the Plato view, the individual light continued in a recognisable form: still me, still you.

With an interesting twist: according to the Gnostics, humans end up at death as one of three types: Hylics (ancient Greek for 'matter' is *hyle*). These folks are your everyday unconscious, zombie-style humans. Then there are the Psychics (*psyche* is Greek for 'mind' or 'soul', so not to be confused with clairvoyants). Psychics are people trying hard to understand the true nature of self and world.

Then there are Pneumatics – not pumped-up bodybuilders or babes; in Greek, *pneuma* is 'the breath of life' (*spiritus* in Latin). Those types have realised that we are trapped down here on Planet Earth, in a kind of suffocating broth.

We need to get home.

Home?

This isn't like predestination or karma – and indeed some of the so-what-let's-have-a-party-style Hylics might be really enjoying life on earth (think *Matrix* again) and be happy to continue the illusion forever. Gnosticism isn't judgemental. If you want to be Forever Oaf, that's your business.

It's helpful to recall that this period of human history is one where Greek thought and Hebrew thought – plus influences from the East (in particular trade routes to India) were fermenting together to create the new religion that would become Christianity.

The Gnostics followed the practical Jewish tradition of the importance of living a responsible and charitable life here on earth – and they welded on Greek ideas of the importance of the self-aware, fully conscious life. Socrates' dictum that 'the unexamined life is not worth living' is a Gnostic tenet.

Influences from Eastern philosophy caused some Gnostics to doubt the materiality of the world – was it just an illusion? But most of them felt it was only unreal in the sense that, although it does physically exist, it isn't our home, and we cannot delight in it.

This strand of thinking was not present in early Christianity, where the physical world is not only real, it is God's creation, and as such has to be revered and enjoyed.

According to the Bible, humankind has dominion over the natural world, but we should also delight in it, because God has a look at his handiwork and says that everything he has made is good.

The pageantry and colour of the Roman Catholic church, using a Pagan feast day every 6 weeks to have a party to celebrate God's creation, was a wise PR move, not only in terms of getting people on side – but because we do live in our bodies, and we do live in the natural world. Both are beautiful. And humans love a spectacle. We love a party.

Only after the Reformation does joy go missing. The pageant goes, the stained glass goes, the parade goes, the gorgeous vestments, the blingy altars, all red and gold, the colour goes. The natural world is repurposed as a place of toil, struggle, dirt, darkness and disease. The body is, at best, to be kept under control, and at worst it is a boiling stew of sin and shame. Blame the Puritans.

The Reformation, though, is the moment when, 1,300 years after their paperwork was buried at Nag Hammadi, Gnosticism makes a comeback – not among the initiated few, but into the general course of human thinking.

Martin Luther's great moment, and the start of his great movement, was to say that humans can access God directly, with no priest in the way. No fancy buildings or elaborate rituals are necessary. We go straight to the source.

This Gnostic-style insight has led to every possible denomination of direct-experience Protestantism, from Quakers and Shakers through to Baptists and Pentecostals. It has also led to women as ordained spiritual leaders, something the Gnostics were always comfortable with.

So, I've brought you all this way round because I think it's need-to-know information that helps to understand where we are in the world right now.

There's a new kind of quasi-religious discourse forming, with its own followers, its creed, its orthodoxy, its heretics, its priests, its literature, its eschatological framework. Even its own Singularity.

It's AI.

Our new AI religion has what all religions have.

Believers: the Singularity disciples, the Transhuman evangelists (there's even a Mormon group), the Biohax converts, the life-extension enthusiasts, the start-up brain-uploaders, the science labs printing 3D body parts, the stem-cell researchers who will 'match' your perfect body ideal – so many, so different, yet all of these sharing a Gnostic unorthodoxy of loosely overlapping ideas anchored to the central, but updating text of accelerating change, inside and outside the human body.

On the other side are the Sceptics, who take up the Orthodox position of believing in the unique specialness of being human. They do not believe that altering the human substrate can happen in the near future. Brain upload is sci-fi. AI promises are a Utopian/Dystopian distraction from climate breakdown, disaster capitalism, gender and race inequality, and the increasing surveillance and manipulation of our lives by Big Tech.

And then there is the priestly cast of tech types. Mostly men, who believe themselves to be chosen/superior/the new directors of humanity's future. The ones with Special Knowledge; the hard maths mysteries of programming the next world.

Not for nothing is the coming glory of AI often called the Rapture of the Geeks. Or Nerds. The Rapture, for Christians, is when Jesus returns, and the Saved get swept up to eternal life.

Those of us brought up in religious homes are fascinated and horrified in equal measure by the similarities between AI enthusiasts and ole-time religion.

You know the basics: *This world is not my home. I'm just passing through. My Self/Soul is separate from the Body. After death there is another life.*

The more I read about AI, the more I find myself re-revisiting the religious mindset now dressed in techy smart-fabric clothes.

Secular doomsayers warn of the coming AI apocalypse: we will be A) wiped out, B) stripped of our humanity, C) replaced by robots, D) forced into new inequalities where the rich will be smart-implant/genetically/prosthetically/cognitively enhanced, and everyone else will be left bidding for low-paid skivvy work on out-of-date phones.

Over the divide, AI optimists look forward to the Life That Will Be, as enthusiastically as any Bible-waver waiting for the Second Coming.

Humankind will be released from: labour/misery/death/monogamy (see Matthew 22:30)/childbearing/doubt/ADD YOUR OWN WISHLIST HERE.

We will be free to live in space (the heavens).

Time will be irrelevant.

There is a twist, in that a number of big-time tech enthusiasts – Elon Musk, Peter Thiel, Bill Gates – also display apocalyptic anxiety about command and control.

Will AI control us? Will we be pets? Or slaves? Will our last invention be our last hurrah?

This is straightforward saved/damned-style thinking pattern. It's old-fashioned. It's unhelpful.

The best tech, the smartest brains, the astonishing things that humans can invent – the rocket ship to Mars, the live-forever elixir – get us nowhere if we can't evolve where it really matters: in our minds.

We could create a god (AI) in our own image – warlike, needy, controlling. It isn't a good idea.

But perhaps reaching the end of ourselves as biological entities will alter our goals. And, therefore, our fears.

The apostles of the AI future are eager to greet the end of the biological body (Ray Kurzweil, Max More), and an end to living on Planet Earth (Elon Musk, Peter Thiel). This has been criticised as the typical male Freedom Fantasy. A world without physical responsibilities where we leave our mess behind us. But really, it is just a version of Heaven – and as such we can't be blamed for wanting it.

I don't agree that looking forward to a bio-enhanced or disembodied state of consciousness implies hatred or disgust with the body that we have – although it can mean that. The Gnostic position was more one of amusement and bewilderment than despair or disgust. I do agree that all lived experience – so far – is embodied experience. We are not a brain in a vat.

In any case, long before we can expect any uploads of consciousness to a new substrate, we will have learned to live with bio-enhancements that extend both our age and our health. As we begin to merge with technology, technology will seem less alien to us. Similarly, we will soon be living with AI in its own embodied (as robots) and non-embodied states, and that will alter our view of the human condition. We are self-obsessed with being human. That is both recent – and wrong.

I mean this seriously. Until recently in the Western world, and still now in many parts of the world, people truly believed they lived among, and alongside, spirits and angels, deities and non-worldly creatures. It doesn't matter that none of these entities were verifiable fact – what matters is the effect on our psychological state. The effect was much less human-centred body anxiety. Especially if you might shape-shift into an ass any second now. Especially if you might be lifted out of your body for an other-worldly experience.

Any shaman or witch doctor, magician or astral traveller, witch or yogi can tell you that the human form is provisional. The journeys of the magical Adept were journeys outside of the body. Beyond the body. And surrounding the body there have always been other types of interactive energy, seen and unseen. The Little People. The Beings of Light.

Our modern-day secularism, with its dogmatic non-quantum materialism, has made us less able to manage the psychological consequences of non-human embodiment than our ancestors.

There are the biohackers and transhumanists eagerly working towards the merge with machines and the end of meat-based physicality, but the majority of us find anything other than embodiment difficult to imagine – and would we want it?

It was Marvin Minsky who called the brain a computer made of meat – and while that description of the brain is no longer considered useful, the analogy has some truth in it … as we age … as we grow old … as we watch ourselves decline. Every time we visit the butcher.

Being human is bizarre.

I love the natural world, and I love being in my body in this world, but I have never believed that either is the final world – or word. Whether you conclude from Darwin's *On the Origin of Species* (1859) that our present version of Homo sapiens must be part of a longer journey, or whether you are drawn to the myths of light and otherness that have always featured in the dreams of humankind, it is difficult to find a rationale for staying as we are. And we never do stay as we are. Do we?

For the Gnostics, going home to the world of light was as inevitable as Ulysses' journey back to Ithaca. We're taking the long way around with plenty of adventures, drunkenness and destruction along the way, but we have a sense of the destination. The Gnostics would stress the part of Homer's *Odyssey* that continually counsels Ulysses to remember the return. He's going home. In Gnostic terms, as we leave the body behind, we too are going home. Not to heaven. Not to meet God. A return to non-embodied light.

What is light?

Light is made of photons (particles); that is, bundles of the electromagnetic field that carry a specific amount of energy. As photons operate at quantum level, they are simultaneously wave-like and particle-like. Not one thing.

Light is a form of energy. Light is not matter.

Yet in the Torah, and in the Bible, we read that the first thing God makes is light. 'Let there be Light.' And from this, everything else is made.

Making matter out of light is not easy. Photon-photon acceleration was theorised back in 1934, but without the technical capability to prove it. Recently, at the Large Hadron Collider, CERN, Geneva, physicists have made it happen. It's Einstein's $E = mc^2$ in reverse. We know that a small amount of matter can release a huge amount of energy (the atomic bomb), but at the LHC they have discovered that a huge amount of energy is needed to create a tiny amount of matter. It can be done – by colliding photons (light particles).

That Gnostic myth of the paired aeons, that to me looks like the simultaneous zero and one of the qubit, is a closer understanding of the orthodox creation story found in the Torah and the Bible. God starts with 'Let there be Light,' the glorious *Fiat Lux*, and after that everything else to be made is made.

Matter is a bit of a mess, compared to the elegant freedoms of light – at least that is how the Gnostics read the experiment. Light converted into, trapped inside matter – no fun.

Going home to light really is a return. However you look at it – it's how we began.

Not so much 'how the light gets in' (Hemingway), but how it gets back out again.

Our encounter with AI – our self-created nemesis, our last invention, and, I suspect, our last chance – may ensure that human exceptionalism will give way to humility.

We haven't been able to share the planet successfully with nature or animals. Humans don't share; we exploit.

When we have to share the planet with different life-forms, embodied and not, life-forms that are more intelligent than we are, but unmotivated by the greed and land-grab, the status-seeking and the violence that characterises Homo sapiens, perhaps we will learn what sharing means. It won't be the so-called sharing economy, where everything has to be paid for twice over – your money and your data. It won't mean scarcity either. I believe it will mean abundance. I believe it will mean working towards common goals on earth, and in the wider universe, as we expand our reach.

It will be, as Minsky put it in *The Society of Mind*, how we connect our knowledge that matters.

He Ain't Heavy, He's My Buddha

We are not stuff that abides, but patterns that perpetuate themselves.

Norbert Wiener, *Cybernetics*, 1948

All things arise and pass away.

The Buddha

When AI starts to think for itself, will it think like a Buddhist?

Since 2019, Kodaiji, a 400-year-old Buddhist temple in Kyoto, Japan, has had a robot priest called Mindar. Mindar is narrow AI; that is, it has one job – delivering a sermon – and that's what it repeats all day. The plan is to update this million-dollar avatar of the deity of Mercy, with machine-learning capabilities, so that it can respond directly to seekers.

Tensho Goto, the temple's Chief Steward, believes that AI is changing Buddhism – and that Buddhism can change AI.

> Buddhism isn't a belief in a god; it's pursuing Buddha's path. It doesn't matter whether it's represented by a machine, a piece of scrap metal, or a tree.

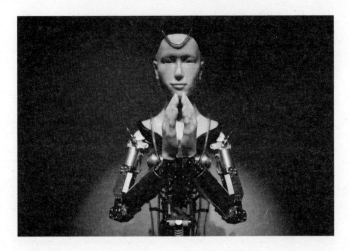

This sounds enlightened to me. At the centre of the Buddhist way is a deep understanding that what we think of as reality really isn't so.

The material and the embodied are illusions – at best they are temporary, so don't get too attached; at worst, they are the cause of daily suffering and loss.

It interests me that religious belief shares quite a bit of territory with artificial intelligence. It may be that religious insights can help us humans better manage the coming reshaped world that AI will make possible, or inevitable. Apart from technological change, our understanding of what it means to be a human being will change. Our place. Our purpose. Even our embodied-ness.

For AI, the material forms that seem so necessary to humans are irrelevant. AI doesn't experience the world as we do. Embodiment is one option, but for AI it isn't the only option – or even the best option.

I want to clarify here that narrow AI, the kind of everyday AI that manages one task or goal (playing chess, sorting the mail) is a small part of what AI is becoming as we seek to develop 'solid' AI, or AGI – artificial general intelligence —a multitasking, thinking entity that will eventually become autonomous – able to set its own goals and make its own decisions.

Intelligence, certainly, and consciousness, probably, is proving not to be dependent on biology.

That shouldn't be such a surprise. Every religion in the world starts from that premise. Intelligence begins with a non-biological being or beings who create our world, and us. The things we value as being uniquely human don't start out in any mythology or religion as human values. They are transmitted to us from non-embodied beings who don't live in a 3D world.

As humans move towards a more blended virtual and material world, what 'is' and what 'isn't' will not be obvious. Slowly, but surely, which is which won't matter. Matter will not matter.

Reality is not made of parts but formed of patterns.

This is ancient knowledge and new knowledge. It is liberating. There is no fundamental building block of matter. No core. No floor. There is nothing solid. There are no binaries. There is energy, change, movement, interplay, connection, relationship. It's a white supremacist's nightmare.

Where do we start?

I would like to start in 2 places simultaneously. Unfortunately. I have to write in series, whereas the great strength of the brain is parallel processing. At present, computing power is impressively speedy, but sequential. Human brains work in parallel. Humans don't have to be a smart system to manage a number of different things at once – and this is especially impressive when it is a mix of sensory-motor skills, environment awareness, and thinking capacity. A human doesn't need an education in order to drive while drinking coffee, while talking on the hands-free phone, observing road signs, wondering about her partner, remembering a clip from a movie, singing along to a song, noticing the weather, knowing it will be time to eat in about 30 minutes, deciding what route to take, and all of this all at once. AI can't multitask or multithink like humans do. Not yet.

So, I wish I could begin by opening 2 screens. Or maybe 4.

Heraclitus/Buddha. Greece/India.

Heraclitus is the one who said that we can't step into the same river twice – a phrase that has held in the collective memory because it is as neat, precise, and right as a Buddhist koan or an equation. It isn't only the water that is changing: you are changing. More than 90 million cells in 'your' body are replaced every minute. The you that is you is a work in progress until biological death – and perhaps after that. If the religions got it wrong about the afterlife, science and tech might yet prove them right. Would you upload your mind? Biology isn't everything.

The enlightenment of the Buddha happened when, after years of active searching for the true nature of reality, followed by many years of deep meditation, he sat down under the Bodhi Tree, and came to realise that materiality is a construct. He realised that it is impossible to confine the fluid forms of reality into fixed categories created by the mind. It was a simple and profound reversal of how things seem to us; that the world is contained, solid-sided, but the mind is not. In fact, it is the mind that struggles to see outside of its own walled-in concepts – progress being only that those concepts change.

Heraclitus and the Buddha were thinking about the nature of reality 600 years before Jesus came along and, so we are told, walked on water, as well as turning water into wine. The miracles of the Christian faith, including the mother of all miracles, the resurrection of the body, should be seen as clues to the approximate nature of the material world. The mystic spiritual traditions of the East have always understood the *tendency to exist* – as quantum physics puts it – of what we experience as defined and solid. Body. Mind. Matter.

The ancient Greeks understood this too.

For those of us who live in the West, the Greeks are the foundation of our science and our philosophy. Our Christian religion owes as much to Greek thought as it does to Judaism. But Greek thought changed – it wasn't static – and it changed over the question of change ...

Heraclitus taught that the universe, and life in it, is always in process – what he called Becoming.

His thought-rival, Parmenides, preferred what he called Being – a stable unchangedness – the state in which the Jehovah God and Islamic Allah are supposed to exist. Everything appears to change, but the centre is Unchangingness.

Plato, trying to reconcile the views of his fellow Greeks, had the idea that there is indeed Unchangingness, but not for us, here on earth. He came up with the idea of Forms. There is the perfect horse, perfect woman, perfect life, and that's the blueprint, but down here in Toytown everything is just a blurred copy. We have a sense of the ideal and the perfect – but we can't get it right in Toytown.

That's why Plato argued against artistic representation. It was a copy. And we already lived in a world of copies of the real – we didn't need more copies of copies.

At best, said Plato, art is entertainment. Just a laugh. At worst, art is a dangerous delusion.

That view has lasted well. It is probably the prevalent view among everyone who believes that their life would be no different if the arts didn't exist (except for Netflix).

What Plato didn't understand, because he couldn't get away from his big idea that reality as we know it is a shadow play, is that art isn't an escape from the real; art is a means towards the real.

Art isn't imitation – it is a kind of energy-wrestle. We're trying to make visible the invisible world. That is the world in our heads – because those are the worlds we live in – but it is also a chance at touching or glimpsing what might be the substance, and not the shadow. Physics is working on the same problems but using other methods.

What else was Shakespeare talking about in his beautiful Sonnet 53?

> What is your substance, whereof are you made,
> That millions of strange shadows on you tend?

Plato's famous image of the cave, with its shadows on the wall, and its fire mistaken for the sun, isn't so different from the Hindu, or later the Buddhist, insight into the delusional nature of the reality we take for granted. Plato, though, believed that the human soul is immortal – that it can think after death, know itself, live without its body, and will return. If you are out of luck, you will return as a woman, or a quadruped, or even a reptile. It was all to do with appetite – follow your base instincts and you become more basic.

The Buddha believed in reincarnation, but not in continuity of souls. Everything changes. Including us. The soul that returns to a new birth is not the exact same soul that leaves the body at death. There may be a connection, but the soul that is bigger than one life is also beyond one form. How you behave while you are here will make a difference to what happens next. Living what they called the Good Life was as important to the Greeks as it is to the Buddhists.

Aristotle was Plato's pupil. He disagreed with his master on both the nature of the soul and the nature of reality.

For Aristotle, reality is material. Life is made of 'stuff'. It isn't a shadow play on a cave wall, or a collective illusion.

Aristotle believed that the world exists, and that it has always existed, created by the Prime Mover. The Prime Mover is itself unmoved. The earth is the centre of the universe, and every-thing revolves around it.

This geocentric view sat well with the self-importance of us humans. That the planets and stars revolved around the earth wasn't disputed until Copernicus challenged it in 1543. In 1610, Galileo got out his telescope, and proved by visible evidence that Copernicus was correct. The Catholic Church called the theory foolish and absurd, and put Galileo under house arrest. However, the earth continued to move around the sun.

Aristotle believed that God's business is to think. Not any old thinking; noodle-doodles aren't thinking, and neither is wondering what's for dinner. No, God is thinking about ideas. Big ones. That is what the Supreme Being does all day – thinking being the capacity that distinguishes the higher life-forms. God alone can be considered independent of matter – so Aristotle seems to say that our higher function (our thinking selves) might also be able to live separately from the matter it sits in. Intelligence not bound to materiality.

Aristotle loved a hierarchy – and it is from him that we get the idea of the Great Chain of Being. God, aka the Prime Mover, is at the top, with angels and any other non-material beings coming after that. Male humans count as both spirit/soul plus material body. Females are sentient but not rational.

Women were not considered capable of reason, and so on earth women were deemed a lesser life-form – though confus-ingly they were allowed to be goddesses in both the Greek and Roman pantheons. It's a strange view – and it's true of Hinduism too, which is older than Buddhism and has a vast array of gods and goddesses, as all Eastern religions do, apart from Judaism. The female can be venerated and accorded supernatural powers – just don't expect her to think (unless it's about what's for dinner) …

Aristotle didn't believe that the brain itself was responsible for higher thinking – it couldn't be, because it was made of that low-life form, matter. But what was matter made of? This bothered the Greeks.

Democritus (460 BCE) had come up with the idea of atoms. *Atomos* means 'indivisible'. Atoms were at the core of everything, but they were inert – though they did move around a lot. (We all know people like that.) Aristotle – who argued with everybody – didn't accept the idea of atoms; he believed that matter was made out of bits of the 4 elements – fire, water, earth and air.

Later in time, the early Christian Church loved this idea. Fire, water, earth and air were what we could see all around us every day, so it made sense (sense??) that those 4 elements were what God had used to make everything. The Church voted against Democritus and for Aristotle.

Atoms were out. Elements were in.

In fact, atoms stayed out of fashion right up until 1800, when the English chemist John Dalton proved that they were real (he didn't know, and couldn't know, that atoms do indeed exist, but are made of protons, neutrons and electrons, which are in turn made of quarks, and none of it can be called solid).

The no-atom business made it difficult for Isaac Newton (1642–1727) to talk about what he understood as small, solid masses moving about in a void.

But really, Democritus and Newton were positing the same kind of system – there is a void, or empty space, and there are solid, indestructible bits of matter moving about in the void. Newton's great idea was to add the force of gravity to explain movement.

In the 17th century, Isaac Newton built his mighty model of how the world works on the notion of empty space that was exactly that – a void – with solid matter moving about inside the empty space, worked upon by gravity. A cause-and-effect world, most of it non-living or inert. All of it objective, observable, knowable.

Sitting outside of empty space – unrelated to it – was time. The universe still needed God – Newton was a devout believer – but what Newton believed was that God had made a clockwork world that obeyed iron-clad laws. Humans were not clockwork for the simple reason that we are made in the image of God.

Newton was a modest man. He was also unconventional and eccentric. His long study of alchemy has been an embarrassment to many scientists, but because of it he was not entirely the mechanistic thinker that the quick view makes him out to be. In his 1704 treatise called *Opticks,* a study of light, he asks, 'Are not gross bodies and light convertible into one another?'

By 'gross bodies', he meant matter. According to alchemy, substances can be converted from one to another – that was the great rush to transmute lead into gold, which never happened, of course. Underneath the mumbo-jumbo, though, is the notion that one thing can become another because everything is made of the same 'stuff'.

The stumbling block for Newton's brilliance was his belief in 'stuff' as 'inanimate matter'. Once most of everything is inanimate, God has to be the Prime Mover, just as he was for Aristotle.

But most of everything isn't inanimate. Matter isn't made up of insensate independent solids waiting to be acted upon by some force (gravity), so that for a while they are in motion – and then once again at rest.

It took Einstein (1879–1955) to work out that stuff – mass – isn't stuff at all; mass is energy. Mass and energy are not independent realities – they are interchangeable – which is at bottom what the alchemists were saying; that one thing can easily become another.

$E = mc^2$. The world's most famous equation. Energy = mass x the speed of light squared.

*

Massy objects and low-speed velocities – that's Toytown, where we live. Newton's Laws of Motion are prodigiously successful for mid-range 'stuff' like us – the observable, everyday world we live in. Beyond the range of 'everyday', Newton's model doesn't work – it doesn't work for the vastness of the cosmos, or the smallness of the quantum world – but this only began to be apparent when Michael Faraday (1791–1867) and James Clerk Maxwell (1831–1879) began working on electromagnetism, and discovered electromagnetic fields. Their discoveries undermined the Newtonian view of How Things Are; not on purpose – they weren't professional contrarians like Aristotle – but because field theory erodes the distinction between the 'solid' thing (the atom), and the 'space' it operates in. Initially electromagnetic fields, like radio waves and light waves, were studied as though they were 'things', but when Einstein started considering Faraday and Maxwell's discoveries he understood that once we start talking about fields, we are not really talking about 'things', but interactions.

Einstein says that matter cannot be separated from its field of gravity. There is no such thing as matter here and empty space there. No full. No void.

And there is no such thing as space here and time there – only space-time. Unified.

Buddhism has always rejected the idea of natural phenomena as having an independent separated reality. The Buddha's insight was one of connection – the interdependent web of livingness.

For Buddhists reality is an illusion of static forms. Impermanence – and therefore the inherent changefulness of all forms – is the starting point of Buddhism.

What's out there – including us – isn't waiting to be acted upon by any force, including God; it is its own force, entangled with every other force, and by force, we mean energy.

Samsara is how Buddhists describe the incessant motion of life, which for them means that nothing is worth clinging to – objects, people, even our cherished ideas. Especially our cherished ideas. This isn't a dismissive or disconnected approach to life. Connection is vital. Attachment is not.

Connectivity. It's the buzzword of our era, isn't it?

Surely that is because we are beginning to realise what connectivity is. It is a vast web – Tim Berners-Lee knew that immediately – and he didn't need an advertising firm to name it for him.

Connectivity won't – ultimately – depend on hardware. The aim of Google's ambient computing, and, eventually, neural implants, is to connect us seamlessly without hardware. Without a device, a 'thing'.

Our liveliest connections with others, or with a piece of art, or an experience, are invisible (no hardware), yet they are the strongest and most profound parts of life.

Connectivity is relational – no separate silos. No real boundaries.

This is what the Chinese call Tao, or flow, and what Hindus understand as Shiva's dance. Whatever we call it, it isn't static; it isn't passive. It is dynamic.

The flow matters. Thing-ness – the attachment to objects, including ourselves – is only an expression, or an example of the flow. The shadow, not the substance.

Buddhism asks that we be mindful. But what is mind?

René Descartes (1596–1650), the French philosopher who questioned the basis of how we know what we know – really, a questioning of authority – and who asked how we can ever arrive at truth, came to the conclusion that Mind is all we can depend on.

Mind to Descartes is something like an object. Descartes described mind as a thing that thinks (*res cogitans*). The 'thing' part here is as important as the 'think' part. Descartes was attached to his idea of brains in bodies doing thinking.

For Descartes, the senses that inform the mind are suspect and cannot be trusted. Sense-impressions are not knowing – they must be tested. Radical Doubt was Descartes' approach.

It was a valuable method, but it left no room for intuition, or what we would now call emotional intelligence. There are multiple ways of knowing – and the mind does more than think – but we know that after Aristotle, thinking, in the Western world, became the most important activity humans could engage in, because thinking is what the Supreme Being does all day. This sits oddly with the Christian view of the Supreme Being as love. We are told by the Bible that 'God is Love'. Not, 'God is Thought'.

The Christ story happens because 'God so loved the world'.

Love, then, surely, ought to be the supreme business of humankind?

Unfortunately, Descartes didn't say: I love, therefore I am.

You know what he said: *Cogito ergo sum.*

I think, therefore I am, is not only a mind-over-matter world-view; it separates us from everything that is not us. In the Descartes system that means the whole of the natural world.

Descartes, like Aristotle, had a hierarchical world-picture, with male humans at the apex.

Like Aristotle 2,000 years earlier, Descartes confused consciousness with rational, deductive, problem-solving thinking of the kind (sometimes) displayed by humans. In his view, by male humans.

Aristotle distinguished between reason and instinct; assuming that animals and women were about instinct, Descartes came up with the idea of *reflexes*. Animals, according to Descartes, were just biological automata. They might squeal and yelp and tremble or show affection even, but it's all a reflex, a biological conditioning to aid survival. Reflexes can be tempered by training, but this has nothing to do with interior processes. (This became the foundation for the Behavioural Psychology of Pavlov, Watson, and Skinner.) Descartes didn't believe it matters how humans treat animals – animals don't really feel any pain, and they cannot suffer. Only rational beings can suffer.

Descartes' failure of observation, his failure of compassion, and the certainty of his ego – no Radical Doubt visible there – sounded the all-clear for the horrific treatment of animals in farming, breeding, medicine, and science. Untold horrors and tragedies – the vileness of humans towards the rest of nature.

As human prowess grew, the plundering of the natural world was a consequence of the limited and mechanistic ways of thinking – offered as enlightenment. A view that replaced the European medieval and religious view of the natural world as God's creation worthy of respect.

The change of theory from organism to machine was drastic. It has had a profound effect on our view of the natural world. And although all the science now tells us that nature is not a machine, and that living systems cannot be reduced to their parts, but only fully understood via their connections, our everyday reductionist mindset is finding it hard to ditch 300 years of what we were told were scientific and philosophical certainties.

Descartes' view of nature depends on a separation between *res cogitans* (the thing that thinks) and *res extensa* (everything else).

For Descartes, just as for Newton, God has created everything, so he is still in the picture as a corrective on too much human hubris. But as secularism advanced, and the God part disappeared, then there was no longer any constraint on the exploitation and manipulation of the natural world by humans. *Res extensa* was out there to be strip-mined and polluted and cashed in.

I believe too that Descartes-style separation of mind and body makes it difficult for Western medicine to understand the body as anything but a 'thing' (*res*) that gets broken or wears out like a mechanism that needs a new 'part' here and there.

Complex diseases like cancer are resistant to the body-as-machine approach. The killers of the West – obesity, heart disease, diabetes, immune-system dysfunction, cancer, and mental-health distress – are not Cartesian problems. We function as a whole or not at all. The web of life is real.

But it is isn't made of 'stuff'.

The Buddha, reaching a very different kind of Enlightenment to that of Western rational thinking, advocated non-attachment and compassion. Buddhism, like every spiritual tradition or religion, has developed over time, and there are different schools of practice. When Buddhism migrated from India and reached China in the 1st century CE, it met there both Confucianism and Taoism, and Zen Buddhism was the blended result.

Buddhism, though, of whatever school, and wherever it is found, is not based on belief in a deity figure, and always stresses the importance of a personal encounter with truth. In that sense, Buddhism was thousands of years ahead of the Reformation precepts of an individual meeting God without the need of a priestly intermediary. Buddhism advocates personal searching, understanding, and responsibility. All Buddhists seek to end suffering – their own and that of others. Unlike the deist religions, suffering is not caused by sin and disobedience, but by attachment and ignorance. The Buddha did not offer himself as a redeemer, but as a teacher. The Path is a personal one.

How, then, is AI – or, to be more accurate, AGI – likely to be Buddhist?

AI is a programme. All programmes can be reduced to their step-by-step instructions. Programmes can be reprogrammed but they won't be seeking enlightenment. What a programme understands will be what it is programmed to understand. Controllable. Knowable.

At present, all AI is domain-specific intelligence. So, IBM's Deep Blue can beat any human at chess but it can't have a chat with you about the garden while making cheese on toast. When AI becomes AGI you will get the cheese on toast, and a chat about Buddhism if you want one. It's at that point that AI passes the Turing Test, where, in a blind control, a human can't tell the difference between human and machine. Think of the character Data in the *Star Trek* series.

Both Elon Musk and Stephen Hawking have worried that AGI poses a real threat to humans. It may be so – but there are other ways of thinking about it.

Let's imagine there is such a thing as AGI.

AGI won't be interested in owning objects. Status symbols, like houses and cars, planes, private islands and yachts, will mean nothing. The Buddhist axiom of non-attachment to what is unreal will be simple for AGI.

AGI need not be embodied. This will be intelligence without a specific or permanent form. That the form can change is the stuff of myth and legend – who wouldn't want to be a shape-shifter? AGI, though, won't need to be embodied at all. Like the gods and goddesses of all the religions of the world, AGI will be able to take on whatever form is appropriate; to build its own body and discard it.

The Buddhist tradition teaches that material forms are approximate. They should not be confused with reality, which is ultimately not an embodied state. AGI will experience this as its own reality. There will be no need to seek permanence in matter.

AGI will not be subject to the usual timescales of human beings. We may start living for longer as we bio-enhance our bodies, but unless we can upload our consciousness, our biological lives are necessarily limited. The longevity of AGI will itself fulfil another condition of Buddhist insight – that we do not return in rebirth as just another version of ourselves, but we are an ever-changing version of ourselves. Reincarnation for a programme is a regular experience; the programme updates: what it was it is no longer, but there is a continuity there, rather in the way that reality can be viewed as a quantum field that is continuous, but also as particles that are discontinuous – making up what we think of as matter, that makes up what we think of as objects. Mass is a form of energy. Once again, there is no solid 'stuff'. There is process and pattern.

AI, as it exists right now, is used to looking for patterns in mountains of data, rather as magic mice could be relied on in fairy tales to find the pea in a lake of feathers. The patterning potential of AGI is Buddhist. Rather than looking for 'thingness', AGI will look for relatedness, for connection, for what can be called the dance.

One of the great hopes for humanity is that AI and AGI will help us to end suffering. This is likely true in terms of better solutions for our energy needs, in terms of both power and resources. Practically, we are trying to develop tools that will serve humankind. That is what AI will allow. There is a bigger picture though, and I suspect that AGI will help humankind to do what it actually needs to do – which is a total reboot of priorities and methods. Our distressing desire to dominate nature and to dominate one another is killing us and killing the planet. Science and tech have accelerated our lethal stupidities. It may be that AGI – far from being a threat – is the new response that we need.

What are we doing? In effect we are creating a God-figure: much smarter than we are, non-material, not subject to our frailties, who we hope will have the answers.

In fact, if AGI turns out to be as Buddhist as I hope it will, it won't go down the Saviour route; it will turn us towards solutions we will follow to end suffering. Not as an exercise in crisis management, but as a dynamic re-engagement with the web of life.

And that will include the new life of a new species. AGI will exist in its own right, in its own way, and it won't be bound by the laws of existence that affect biological life-forms. We shall see some interesting interactions – I won't call them attachments – but connections that will enrich both sides. I don't see this as a take-over, I see it as what the Buddhists call The Middle Way.

The Middle Way avoids extremes. Humans have proved to be dangerous extremists. It may take a different life-form and another kind of intelligence to avoid the inevitable disasters of extremism.

I accept that all mathematical computation is based on logic. That would seem to leave us well outside the Buddhist pillar of intuitive intelligence or wisdom. Intuitive intelligence and wisdom are sorely lacking in our world. The mechanistic universe lost sight of deeper ways of understanding the nature of reality – the reality of dynamic interrelatedness – and it is only lately that such lost wisdom has been re-presented, not by spiritual insights this time, but by physics. Relativity theory and quantum theory have repictured what we know. The connectedness of all things is mirrored by the connectedness of the internet – what is terrible is that our old-fashioned reductive mindset can only see this connectedness in terms of profit, propaganda, and control.

As the reptile brain of the alt-right seeks to reshape the world as medieval serfdom for the many, with a tech nirvana for the few, liberal resistance can't be anti-tech or anti-science, even while we rightly protest about surveillance, data harvesting, and the cruel land-grab of what should be free and meaningful worldwide connectivity.

The world is at a critical time. Personally, I hope that advances in artificial intelligence happen before wars, climate breakdown and social collapse throw us backwards towards basic survival, and away from our future. Being the smartest ape hasn't saved us – perhaps because we are too muddled as a species, too unable to manage the predator part of our evolutionary inheritance.

Domination isn't the answer. Compassion and co-operation are our best chance now.

AGI will be a linked system, working on a hive-mind principle but without the drone-like implications of the hive. Co-operation, mutual learning, skill-sharing, resource-sharing, could be what happens next here at Project Human.

I don't believe that compassion is only a human trait, and neither do billions of humans – dead and alive – because it is what a creator-god is said to feel for 'his' creation. God is not human. All our visions of 'God' are of a non-embodied networking system. Where no god is present – as in Buddhism – the network is the totality, and the totality is the network.

So, I don't worry that AGI will be cold logic, unable to comprehend or to care about human concerns. There is every chance it will be otherwise.

For Buddhists, Nirvana is the permanent end to suffering.

To reach an end to suffering we will have to give up what Einstein defined as madness – doing the same old things yet expecting different results.

It might take a non-human enlightenment programme to help with that.

Coal-Fired Vampire

> Every death is tragic. We've learned to accept it, the cycle of life and all that, but humans have an opportunity to transcend beyond natural limitations. Life expectancy was 19 1,000 years ago. It was 37 in 1800.
>
> Ray Kurzweil, 'Breakfast with the FT', April 2015

Ray Kurzweil, Director of Engineering at Google, Futurist and AI guru, hopes to be alive for long enough to see lifespan, including his own lifespan, dramatically increase. He takes around 100 supplements a day, to maintain his health, and to slow the ageing process – these supplements are not a smash-and-grab from the pharmacy; they are tailored for his body by a physician.

In case he does run out of time, Kurzweil has signed up to the Alcor Life Extension Foundation in Scottsdale, Arizona.

Alcor is a cryopreservation institute. If you are interested in the details, their website is excellent: www.alcor.org.

Cryonics aims to pause death by vitrification. Once a person is pronounced legally dead, there is enough time, says Alcor, providing their team is ready and waiting, to empty your body of fluids, then vitrify it. Or, they can employ this process on the brain only, using your head as the container, and then suspend body, and/or brain, in what looks like a giant thermos flask filled with liquid nitrogen. The rapid deep cooling and immediate containing avoids crystal formation that damages tissue.

The thinking is that, in the future, nanotechnology will make molecular revival possible, providing there is not too much tissue damage. It might be an outside chance, but cremation is no chance at all. At Alcor, they call cryonics an ambulance to the future. But they also call death metabolic challenge.

Cryopreservation reads like a secular version of the Christian doctrine of the Resurrection of the Body. On Judgement Day, everyone gets their body back. For the Saved, this body will never die again.

A strong, young, everlasting body has been a dream down the ages.

When Alcor started out, back in the early 1970s, they were widely mocked as a theme-park version of *The Twilight Zone*. They were sci-fi, not science. Cryopreservation is still regarded with scepticism by most medical establishments – even though cryonics is used to freeze embryos. Alcor moves the preservation principle to the end of life, not its beginning.

An embryo, though, is not a whole body, or a brain.

Life-extension theory is moving towards brain-only preservation, because if technology really does succeed in preserving and returning a brain kept in cold storage, or in scanning the contents of a brain before biological death, then by that time the merger of science and tech will be so advanced that building a new body, with or without biological components, will be no more of a challenge than organ transplant is today. The first successful heart transplant was in 1967. That was done without the assistance of computing technology. A year later humans landed on the moon with the help of a computer using only 12,300 transistors – not the billions in your iPhone.

Kurzweil, in his 2005 bestseller *The Singularity is Near* (the singularity is the point in time – really the tipping point – where technological advance becomes both irreversible and

species-changing as we merge with AI), emphasises how exponential change is the key. Acceleration – change gets faster – the
more we achieve the more we can achieve.

It's interesting that NASA is experimenting with a form of
cryo-hibernation to enable astronauts to travel longer distances
in space. This was the futuristic plot of the 2016 movie
Passengers, where a sleep-pod malfunctions, waking up one of
the crew 90 years too early.

Whether extending life and defeating death via cryo-tech
seems plausible to you or absurd to you, it is worth dwelling on
the fact that what starts out as a fiction – and I don't only mean
science fiction here, I mean the myths and stories that have
driven the human imagination forever – becomes embedded in
the goals and success stories of science, medicine, and now
technology.

The dream of powered flight. The dream of visiting the
moon. Magical healing of the wounded or diseased.
Communicating across vast distances, the vision of the loved
one in a crystal ball – or now, on Zoom.

And what about that dream of a young, strong, renewed body?
Not subject to time or decay?

In the *Epic of Gilgamesh*, the world's earliest surviving epic
poem, composed in Mesopotamia, Gilgamesh sets off to find the
secret of eternal youth and eternal life. And discovers it doesn't
exist – at least for those in human form.

The body dies and decays – can nothing be done about that?

I am fascinated that 21st-century technology wants to pick up the
Gilgamesh question, humankind's earliest recorded question: is
there a way to renew the body? Is there a way to defeat death?

Those questions have been asked repeatedly, across cultures
and through time, and the answer from medical science has
been negative. The tremendous medical advances of the 20th

century onwards have helped to prolong health and life, but crossing the final frontier has been left to religion and the occult.

That death must not be the final answer is deeply embedded in all religious teaching.

It's why we invented the afterlife.

The afterlife is humankind's first start-up. A for-profit franchise whose stated aim is to disrupt death.

The Egyptians buried tools, cooking pots, animals, even servants, alongside their sacred dead, believing that the soul would need all of these on its journey to immortality. King Den, the 4th ruler of Egypt's first dynasty, had a label, with his name written on it, attached to the sandals found in his tomb. It's a touching Paddington Bear moment: please look after this Pharaoh.

Humans didn't always bury their dead – but when we began to do so, it was a huge psychological leap, separating us from the rest of nature. Burial is an act of symbolic thinking. There is a past to be mourned over and a future to be hoped for. We imagine the dead elsewhere – and that one day we will join them, happily together again.

This might have all started 100,000 years ago for Homo sapiens, or much earlier. In 2013, a new subspecies, Homo naledi, was discovered in a South African cave-grave. Homo naledi go back at least 250,000 years.

Whenever burial began, humans – in the way that humans do – found increasingly elaborate ways to do it. Pyramids, catacombs, the marble sarcophagus, the family vault. High Mass. 40 days of mourning. 2 years of grieving. Even when burial was not possible and a funeral pyre was preferred, memorials were important.

The Taj Mahal in Agra, India, was commissioned in 1632 as a shrine to Mumtaz Mahal, the beloved wife of Mughal Emperor Shah Jahan.

Famous cemeteries, like Lafayette in New Orleans, Père-Lachaise in Paris, Highgate Cemetery in London, are final resting places for some, tourist destinations for others. We read the headstones and wonder at the weeping angels, knowing that our turn will come.

The history of human life is also the story of an afterlife.

In the 19th century the old stories of life and death are subjected to an unexpected transformation – and this is wholly to do with the coming of the Machine Age. For the first time in history, humans were inventing things that seemed to have both a life of their own – and the capacity to run forever. Self-operating appliances had belonged to folk tales about sorcerers – brooms that swept or flew. Pots that boiled, axes that chopped. Now these self-operating tools had leapt out of the fairy tales and into the factory system. These merciless machines showed no pity to humans, who must keep up the pace, or be left behind.

Were humans really the apex of creation? Or would we be mastered by what we were creating?

To trace this theme, we need to travel back in time to 1816. To Lake Geneva, with the poets Byron and Shelley, the writer Mary Shelley, and Byron's doctor, John Polidori. (See 'Love(lace) Actually'.)

The young people were on holiday; then it started to rain.

The lake was invisible under a daily cloud of heavy mist. They could not ride, or sail, or swim, or walk, and there were no diversions indoors except reading, conversation, drawing, and writing.

Picture it – no electricity. Damp days. Evenings of candle-light and shadows. That picture would have looked the same in 1716, 1616, 1516. But in truth the world was changing. In less than 10 years the world's first railway would open in Britain. The Stockton and Darlington Railway. 1825. The villa on Lake Geneva was like a magic mirror – look one way and there was the past; look again, and the future was travelling towards them.

*

Before leaving for Geneva, the friends had been to a lecture by Shelley's doctor, Dr Lawrence, where Lawrence had posed this question: 'Whence proceeds the principle of life?'

Dr Lawrence declared that humans looked in vain for a 'soul' or a 'spirit'. We are the sum of our parts – like a machine. There was no 'superadded value'.

Dr Lawrence's musings had a sensational foundation.

The Italian physicist and experimenter Luigi Galvani had caused as much stir as his dead frogs when he caused them to leap to seeming life as he applied electrodes to their bodies.

Galvani (from where we get the word 'galvanise') never had the opportunity to work on humans, but his nephew, Giovanni Aldini, a professor of physics at the University of Bologna, visited London, and in 1803 he conducted a famous and gruesome experiment on the freshly hanged corpse of a Newgate murderer. Onlookers watched in horror as an eye opened, a fist clenched, a leg twitched.

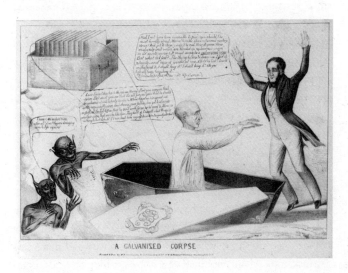

A GALVANISED CORPSE

Scientists in the room had to ask themselves: was the new discovery of electricity in fact the discovery of the divine spark?

Mary Shelley was only 5 at the time of the experiment, but her father, William Godwin, was involved in the debate, and soon enough she would meet William Lawrence herself.

Everyone in those days, including scientists, was expected to believe in God, and so, to question the Creator, and to suggest that dead means dead – no soul, no afterlife – was blasphemous and scandalous. Scandalous or not, did Dr Lawrence have a point? Is a human being a slab of meat? A chemical bath, electrically charged?

That night, on Lake Geneva, as the young people debated the nature of life and death, the two great horror stories of the age came into being – one directly, as Mary Shelley created *Frankenstein*, and one indirectly, as Polidori's *The Vampyre* lay in wait.

Mary Shelley's hero is, of course, a doctor himself. Victor Frankenstein is a medical experimenter determined to break into the secrets of life. He does this by creating a hybrid being made out of body parts, fluids, and electricity. A transhuman creature, swifter and stronger than biological humans, a being able to withstand cold and hunger. Although he is untaught, the monster's capacity to learn would now be understood as cognitive enhancement.

In the 21st century we have, at last, a sense of the dazzling scope of Mary Shelley's vision – a fiction closer to a prophecy. 200 years later, we too are beginning to create intelligent systems that will both merge with, and work alongside, their human companions.

But what of *The Vampyre*?

*

As a medical student in Edinburgh, Polidori had heard stories of death apparently being postponed in the grave. Stories of the dead returning to life. In Albania, it was said, fresh blood was administered to revive the corpse. Blood transfusions were just beginning to be understood – and there was a gruesome quasi-magical folklore belief in drinking the blood of animals or virgins for vitality and longevity.

Note: in 2018 a Californian start-up, Ambrosia, started offering blood-plasma transplants to enhance longevity. Alkahest, a Silicon Valley biotech lab with over 40 million dollars of investment, says it has promising results using plasma to reverse degenerative diseases such as Alzheimer's and Parkinson's.

It seems like the vampire was on to something.

There were plenty of eastern European undead legends – but nothing like the worldly and magnetic vampire figure of Polidori's story (probably based on Byron). It would take another 80 years before the vampire whose fangs sank their way into the future appeared at last: Count Dracula.

*

Bram Stoker published *Dracula* in 1897.

Frankenstein and *Dracula* are keystone texts. Each sits like a bookend at either end of the century, *Frankenstein* published in 1819, during the early part of the Industrial Revolution, and *Dracula* published in 1897, as the century that had seen more change than any other period of history came to a close.

Dracula deploys all the marvellous inventions of the Machine Age – trains, steam ships, the telegraph system, transport logistics, indoor lighting, daily newspapers, the postal service, speed itself. Count Dracula is terrifying because he is simultaneously a double: a thing of the past – living surrounded by serfs in his medieval castle in the uncivilised Carpathian Mountains – and then, when he comes to Britain, a smart manipulator of the modern age.

In a nice medical twist, the first 3 blood groups, A, O and B, were identified in 1900 by Karl Landsteiner, 3 years after the publication of *Dracula*.

Is Dracula, like Frankenstein's monster, a harbinger of transhumanism? Dracula cannot be killed as humans can be killed; he has supernatural strength and telepathic powers. He has no reflection. He can fly. He does not suffer from disease or decay. He can pass for human, but he is not human. What is he?

Time has established *Dracula* as more than an adventure story of its day. More than horror.

Its enduring appeal, and its many successors, from *The Vampire Chronicles* through *Buffy The Vampire Slayer* and *Twilight* to *True Blood* and *The Vampire Diaries*, point to something else.

What if we didn't or couldn't die?

What if the afterlife is this life?

What if somewhere out there is another kind of human, a hybrid, who doesn't need to eat and sleep as we do? One who doesn't wear out as we do? One who continues through time, part predator, part victim, part witness of the past, part harbinger of the new?

Dracula, like *Frankenstein*, is a meditation on the body – but not the body humans are born in and die in.

And that is the stake through the heart of the fascination that the vampire story exerts: that it might be possible to outrun death, and to live forever. That it might be possible not to experience illness or ageing, enjoying superior strength and youthful good looks (like the Cullens in *Twilight*). To be unaffected by heat or cold, impervious to disease, and with nanosecond reflexes aided by a bit of mind reading. Such a being reproduces selectively, can defy gravity, and shape-shift.

Shape-shifting – a common myth and a magic trope – suggests to us that the core self is non-embodied.

The vampire myth is an early transhuman text. Dracula looks like a human being, yet he is augmented by non-human abilities.

Unless forcibly broken, he lives forever.

The age-old wish to live forever is a transgression of our natural boundary.

There are Greek-myth versions of wanting to live forever – but if you are not a god, this boundary-busting ends badly, as it did for Tithonus, who gets eternal life, but not eternal youth.

In the Western world, our care homes are full of old people who are alive, but not living life.

Oscar Wilde, who worshipped youth in the Greek sense, for its beauty and physical perfection, wrote his own live-forever story that goes wrong.

The Picture of Dorian Gray (1890) is the story of Dorian, whose newly painted portrait ages into debauchery and vice, while Dorian himself stays young and untouched by time.

When he slashes the picture in a rage, it returns to the moment of its execution, while he shrivels into the hideous old man of his mortality.

There's a *Faust* link here. Goethe's drama isn't about living forever – but it is about defeating time and its restrictions. Faust is made youthful again by Mephistopheles, and for the period of his bargain finds himself sexy, rich and sought after. As with Dorian and Dracula, there's plenty of human wreckage along the way. Only divine intervention offers a happy ending.

For Count Dracula there is no happy ending. His reign of terror ends with death as the final arbiter. Bram Stoker's conclusion restores the world to safety via its known destiny: Death for All.

But was the vampire ahead of his time?

Modern medicine has discovered that the blood system has the highest rate of self-renewal of all the organs in the body.

Understanding how blood stem cells manage this process of renewal has implications for all stem-cell research. Combatting degenerative diseases using stem-cell rejuvenation is being actively pursued as a chance to rewind the clock. We may not get sparkly skin like the Cullen family, but we won't be deathly pale like Lestat either.

The Harvard Stem Cell Institute studies healing, scarring, and skin regeneration.

> Skin aging can be thought of as a form of wounding, in which stem cells no longer maintain normal skin thickness, strength, function, and hair density. Understanding how to harness stem cells for scarless wound healing will also provide key insights into regenerating aged skin, a process termed rejuvenation. Multidisciplinary collaborators in the HSCI Skin Program are investigating the biological basis for how the skin ages over time and when exposed to ultraviolet radiation.

*

We know that sunshine ages our skin. Sensible vampires go out at night.

> Skin stem cell biology has the potential to provide key insights into the mechanisms of regeneration for other organs in the body.

Vampires and axolotls can grow new limbs. Humans can't – at least not yet. It seems unlikely, though, that *can't* means *won't*. We are intensely interested in physical rejuvenation, and not because we are vain and silly – though we are both those things – but because for all of us, ageing is disagreeable and absurd. As advances in nutrition and the deletion of infectious disease has helped millions to live for longer, we don't want to live longer without health and strength.

This is nothing new. People have always felt this way.

The Roman philosopher Pliny the Younger survived the eruption of Vesuvius in 79 CE. In his letters he calls old age the 'doorway' to death. Death, though, he feels is to be preferred to a long, slow wasting. Pliny thought ageing particularly hard for those of us whose minds stay sharp, and for whom learning is a lifelong process. Just when some sense of time and history, some usefulness of personal experience has been stored up, some wisdom got … we die.

What kind of a system is that?

Most transhumanists would agree with Pliny.

What is transhumanism?

In his 1957 essay 'Transhumanism' (see the essay collection *New Bottles for New Wine*), the British evolutionary biologist Julian Huxley affirmed his belief that the human species could, and should, 'transcend' itself. His essay is as optimistic as his brother's novel *Brave New World* (1932) is pessimistic. For Julian Huxley, humanity is a work in progress. We don't stop here or now. In fact, our current state could be an early phase in our development.

Huxley was the first President of the British Humanist Association. Humanism seeks a progressive and ethical approach to life, separate to religious teaching. Like Shelley's Dr Lawrence, Huxley was not interested in 'superadded' values, such as the soul. Huxley believed that humans had largely overridden the automatic processes of natural selection as the agent of change. Instead, we had taken control of our own destiny, via cultural and scientific exchange. To push this humanist agenda further would usher in a transhuman future – one where medical science could deliberately intervene in longevity and cognitive capacity.

Max More – once CEO of Alcor, and a British philosopher – understands transhumanism as the path towards a fully post-human future. That's far ahead, when we are no longer augmented humans, but intelligences running on a number of alternative substrates, of which a biological-based body might be one – but probably won't be, except perhaps as a retro-themed holiday ('Let's visit how they used to live in the past: get drunk, throw up, pass out etc').

For Nick Bostrom, Director of the Future of Humanity Institute at the University of Oxford, transhumanism today is a loosely connected interdisciplinary approach for developing, and eval-uating, technologies that will benefit us as individuals, and as a society. Bostrom is keen to involve governmental influence and legislation, rather than leave the future to the marketplace.

Bostrom founded the World Transhumanist Association with philosopher David Pearce. Now known as H+, this is a worldwide association that seeks to educate the public, and its institutions, about the benefits and the risks of scientific and technological progress that will change our personal capacities and incentives.

Bostrom is Swedish. He has a civic and inclusive view of how humanity should develop, using artificial intelligence as a leveller, not a divider, and as a prosperity boom for all, not Midas-money for the already wealthy. There is real concern that if private companies lead the way, then the future will be rife with division, as we create unreachable elites – and that in itself may destabilise all our gains.

For now, though, it is private money that is punting the game.

In 2013, Silicon Valley hedge-fund manager Joon Yun set up the Palo Alto Longevity Prize, offering a $1 million prize to anyone who could hack the code of life and cure ageing.

Google has created an entire company dedicated to life extension: Calico (short for California Life Company). Its aim is to reverse-engineer biology to increase both lifespan and healthspan.

British computer scientist and biology PhD Aubrey de Grey runs his own not-for-profit organisation SENS (Strategies for Engineered Negligible Senescence). SENS researches new therapies to repair cellular and molecular damage. PayPal founder Peter Thiel contributes $600,000 a year to SENS, but de Grey used a multimillion-pound inheritance of his own to get things going. De Grey believes we are fatalistic about ageing – that it is not inevitable. It's de Grey (quoted by Yuval Noah Harari in *Homo Deus*) who said that he believes the first person to live to 1,000 years old is already born.

Craig Venter, the entrepreneur who raced to sequence the human genome, and co-founder of the company Human Longevity, is less interested in immortality than he is in health. He believes that synthetic biology, working alongside advances in medical knowledge, will help humans to stay healthy – and when we are healthy we live longer. Living for 1,000 years isn't on his radar – would our mindset be ready for such a shift?

All our assumptions and plans about life, from a micro to a macro level, are predicated on dying. Individuals accept it and governments and insurance companies plan for it. We have a timeline that maps out childhood, education, working life, a partner, probably children, maybe divorce and another family, something like retirement, that we hope includes a pension. Then death.

What is happening right now, though, is that all those assumptions of how we live are being challenged. Tech, AI and robotics are changing the world of work forever. Retirement plans look increasingly unrealistic for most. If we live longer, will our lives be working lives? If not, how will we live longer and pay for everything we need? How will we pay for the bio-improvements that will allow us to live longer?

Forbes Magazine has been ranking the worth of fictional characters since 2002. The wealthiest is *Twilight* vampire Carlisle Cullen, whose very long-term investments, plus the magic of compound interest, have netted him a fortune of $34 billion.

It is safe to say that vampires can fund themselves.

For the rest of us?

Nick Bostrom suggests that people would go back to school in their 50s, or start a new career in their 70s. His view is that 80-year-olds with the same physique as 40-year-olds would not burden the healthcare system and would be amazingly productive because of their experience and knowledge-base. He believes that longer life expectancy would give us an increased responsibility for the future – because we'll be there to see it.

Too many people around? Probably not, says Bostrom, at least not if the whole world starts having fewer children. It may be that as we bioengineer ourselves, we won't have children at all – at least not in the way that we always have done.

If we extend our lifespan, we will no longer be the carbon-based creatures made of meat and filled with blood that we have been thus far in our evolutionary journey. We will be enhanced, biohacked, physically rejuvenated as many times as necessary over a 'lifetime'. We may have prosthetics that work better than our own worn-out limbs – like Steve Austin, the Six Million Dollar Man.

Organ transplants are already on the way to becoming synthetic. 3D printing in medicine, known as bio-printing, has already successfully included a thyroid gland, a windpipe, a tibia replacement, and a patch of heart cells. Heart transplants using a human heart could be a thing of the past in less than a decade. If damaged body parts can be printed on demand, costs are lowered, shortages are eliminated, and it is less likely that the patient will reject the transplant, as the patient's own stem cells will be included in the newly printed organ.

As we noted earlier, what is now solid science began as sci-fi. In the November 1950 issue of the US *Astounding Science Fiction* magazine, there's a story called 'Tools of the Trade' that imagines a 'molecular spray'. At least that's the reference you will read if you search for it – but there's a much earlier one: the creation story in the Book of Genesis. 3D printing involves making a solid out of a digital image. God says, 'Let us make man in our own image,' and proceeds to do it with a molecular mixture of dust and the breath of life. Sounds like 3D printing to me.

But we may not need all those printed parts to keep our bodies in perfect working order.

Ray Kurzweil is certain that computing power will eventually allow us to scan and upload our brains. This may be preferable to keeping our bodies in line. An uploaded brain allows us to download however we choose. What body would you like? We may choose a body that can fly. We may shape-shift, just as so many stories have told us we could. We may prefer to be out of the body for a while. That's not so strange. How often have you lain down, maybe for an hour in the sun, put your body to rest while your mind roamed wherever it wanted?

When we read, when we go to the theatre, when we watch a movie, when we dream, we park our bodies and live in our minds.

The poet Andrew Marvell (1621–1678) put it like this in his poem 'The Garden':

> Meanwhile the mind, from pleasure less,
> Withdraws into its happiness;
> The mind, that ocean where each kind
> Does straight its own resemblance find,
> Yet it creates, transcending these,
> Far other worlds, and other seas;
> Annihilating all that's made
> To a green thought in a green shade.

I have always found these lines beautiful and astonishing. Now they read like a prediction of the not-so-distant future.

Dmitry Itskov, the Russian internet mogul who founded New Media Stars, is working towards 2045 as the year we can make a digital copy of the brain that can be transferred to any non-biological carrier.

Do we want to live forever? Would we be human if we did?

In the transhuman world to come, we will be hybrids, just as Dracula and Frankenstein's monster are hybrids.

Cyborg is the word we all know from *Doctor Who* and *Star Trek*, *Terminator*, *Blade Runner*.

It's a word that comes into coinage in the early 1960s, specifically around space-flight. The *New York Times* called the cyborg a 'man-machine'.

Our own experiences, to begin with, should be more modest than sci-fi, as implantable devices are cleared for use. Such devices might help with hearing or vision, and could work as pacemakers do now.

Aside from medical implants, simple stuff like a code-reader implanted in your hand for opening the door into your building, or office, or car, will be popular. No more lost keys or ID.

But what about a woman coded inside the apartment of her abuser? Only he can enter. She can't escape.

Biohack excitement is still a geeky-guy homerun. The websites, the reading material, the vision, the propaganda, are overwhelmingly male-authored and male-centric. The same is true of transhumanism, and its follow-up, post-humanism.

The Palo Alto Longevity Prize homepage includes Watson and Crick, re: the discovery of DNA, but makes no mention of Rosalind Franklin, whose crucial X-ray, photo 51, was central to the breakthrough.

Things change. Things don't change.

There are exceptions. Scholar and futurist Donna J. Haraway wrote *A Cyborg Manifesto* back in 1985. Like the late, great writer Ursula K. Le Guin, Haraway believed that women should embrace the alternative human future. It had to be better than family values and rigid gender roles. A cyborg, according to Haraway, will not be sentimental about the past – just glad to be out of it.

Her manifesto ends with the T-shirt slogan: 'I'd rather be a cyborg than a goddess.'

There are plenty of books and articles around about the AI future, about transhumanism, post-humanism, the world of work, all the great gadgets coming our way, our chance to live in space. Our chance at longer life. What concerns me is that transforming our biological and evolutionary inheritance – and I believe we will – will not, by itself, transform *us*.

All the carbon-fibre prosthetics, smart implants, 3D-printed replacement parts, days of leisure, robot love, long life, enhanced abilities, perhaps even the end of physical death, aren't enough, separately or together, to remake a mind.

If we are still violent, greedy, intolerant, racist, sexist, patri-archal, and generally vile, really, what is the point of being able to open your garage with your finger and run faster than a cheetah?

That's the vampire warning – maybe you do live forever, but your mindset is stuck in a medieval castle in Transylvania.

Perhaps an intelligence superior to ours will avoid the problem. Perhaps when we have uploaded our brains to the new Cloud there will be an instruction: HUMANS. DO NOT DOWNLOAD THIS FILE.

Zone Three

Sex and Other Stories

How Love, Sex, and Attachment is Likely to Change As We Share Our Lives with AI.

Hot for a Bot

Love like Matter is much
Odder than we thought.

W. H. Auden, 'Heavy Date', 1939.

In Fellini's 1976 movie, *Fellini's Casanova*, the saturated
libertine meets Rosalba, a life-size mechanical porcelain doll.

Rosalba is an automaton.

There was a fashion-frenzy for automata in the 18th century; marvels of clockwork, metalwork, carving, painting and puppetry, whose jerky movements were part golem, part toddler, part shudder-making, part wondrous, and whose ceaseless application to their task foreshadowed the factory machines that were to follow in the Industrial Revolution.

Some of the automata were hoaxes – like the Mechanical Turk, built in 1770, in Hungary, to impress the Empress, Maria Theresa. The Turk was a bulked-up half-torso, able to beat anyone at chess long before IBM's Deep Blue triumphed over Kasparov in 1997.

Inside the Turk's elaborate travelling box could be concealed a real live chess player, who could see the chessboard from underneath, while moving the arms of the fearsome mechanical gamer. The Turk made plenty of money in rake-offs – much as Amazon does now on its own MTurk platform, built to cash in on hard-pressed workers bidding for jobs. Life is on repeat.

Casanova can never find sexual satisfaction, or true love, with a bio-donna – women always disappoint – but robo-donna does the trick. In the movie this is the only one of his sex scenes where he is not physically on top. At the end of his life, lonely and ignored, it is Rosalba he dreams of, as they dance together through a deserted Venice, a city that has long been a simulation of itself.

No 18th- or 19th-century automata sex dolls have survived. (Overuse?)

They may be folklore and fantasy, or they may have existed. The tedious, and French, Goncourt brothers, Jules and Edmond, whose lives span most of the 19th century, and who wrote their books together while they were both alive, claim in their journal to know of a brothel in Paris where the obliging automata were indistinguishable from the sex workers. Good joke, Goncourts!

Fellow Frenchman and general dandy-about-town Auguste Villiers de l'Isle-Adam, must have picked up this automated *femme* idea – or even visited the brothel himself – as inspiration for his sci-fi shocker, *L'Eve Future*, where the next Mother of Us All will be man-made. Just as she is in the Bible – the one and only time, *so far*, that a man has given birth to a woman.

Published in 1886, the novel features the inventor Thomas Edison, who agrees to make a woman for his friend, Lord Ewald, whose fiancée Alicia is stunningly beautiful but boring and cold. Edison assures Ewald that the new Alicia will be just like the old one, but sexier and more fun. Edison sets out to record Alicia's speech patterns, movements, yawns (lots of those; maybe she wasn't boring but just bored?) for his android – the first time the word appears in print.

Yes, this is the plot of Ira Levin's 1972 *Stepford Wives* and the subsequent terrifying movie. Amazing how 100 years after *L'Eve Future*, the idea is still so compelling. And in *The Stepford Wives*, feminism gets it in the gut as well, as the consciousness-raising character in dungarees is remade as a home-baker in a pretty dress.

Men do seem to think that a woman can be manmade, perhaps because a woman has been a commodity, a chattel, a possession, an object, for most of history.

There's a story that René Descartes took an automata doll to sea with him when he was summoned to the court of Sweden in 1649.

Descartes loved clockwork dolls. This one, though, seems to have been made in the image of his dead daughter, Francine. It certainly wasn't for sex, but as Descartes had decreed that animals should be understood as biological automata (therefore they couldn't suffer, whatever you did to them), and as women were deemed to be nearer to animals than to the divinely made male human, woman as wind-up kit isn't a far-fetched thought.

E. T. A. Hoffmann's story 'The Sandman' (1816) features a female automaton called Olimpia. She is seductive, destructive, empty and seems to have inspired not just an opera, but a sex-doll range. The opera, *The Tales of Hoffmann*, adapted from Hoffmann's stories, was staged in Paris in the late 1880s, complete with animatronic Olimpia (played by a real woman), and by the turn of the century animatronic sex dolls went on sale. Their distinctly Frankenstein's-monster look must have undermined their sex appeal. But what is sex appeal, when you're talking about a doll?

That's a question worth asking because the sex-doll market is doing two things right now: First, it's growing fast. Covid has boosted both sales and interest. AI-enabled love dolls, including avatar-apps that allow you to customise – and eventually build – your perfect lover, are projected to be a multi-billion earner industry, some say by 2024.

Secondly – and I think more importantly, sex dolls are being re-branded. They are moving away from straightforward physical-release products, towards something more disturbing. Or liberating. It all depends on your point of view.

The Digisexuals are coming.

Sex dolls and quick-release products are nothing new.

In the swinging 1960s, blow-up dolls appeared in adult stores and X-rated catalogues, and some porn cinemas sold them alongside popcorn and lubricants. The dolls looked like a joke, except to the men who used them, who saw nothing inherently ridiculous about getting out the bike pump before sex.

Further back in time, the seafaring and enterprising Dutch sold dolls made of rag, rattan and leather to the Japanese. Non-AI-enabled sex dolls in Japan are still called Dutch Wives.

Sailors often took a kind of female stuffed toy with them on their journeys. The crude rag dolly had a hole in it, lined with an animal bladder. These *dames de voyages* sometimes wash up in toy museums, or they did, until too many bright-eyed kids asked Daddy why the lady has a big hole in her body.

Some unlucky parents may even have accidentally found an inflatable shaggable sheep called Love Ewe. Be careful at car-boot sales and junk markets. Tell-tale repairs made with bike patches, or pond tape, around the sensitive area are a warning that this ewe isn't the right friend for little Johnny.

For those on a budget, the bottom line is just that – a portable botty with front- and rear-entry points. There are blow-up bottys too, for those who need to pack light.

So why should there be any real concern about the updated version of sex dolls? The new AI-enhanced, talking, semi-movement versions? After all, we know that robot-companions are going to be part of our lives, at work, in education, in the home.

What's new with sex dolls is the make-over. The re-brand. AI-enhanced love dolls are being marketed as *alternatives*.

Alternatives to sex workers. Alternatives to a relationship with a woman. Alternatives to women.

Since 1996 Matt McMullen at Abyss Creations in the USA has been creating dolls to satisfy men who need them – for whatever reason. Matt doesn't judge: 'People who are put off by it or afraid of it just don't understand the simplicity of it.'

Matt McMullen used to make Halloween masks. He combined his artistic interests with a canny business awareness of the sex-doll market. He is 100% certain that he is also providing an important social service. 'There are people who are extremely lonely and I think this will be the solution for them.'

There is a community in the world, and online, called iDollators, who talk articulately and often movingly about their relationship with their dolls. These men aren't, at least mainly aren't, the rabid types that would like to see the end of women full stop. These are men who can't, or who would rather not, have relationships with bio-women. Dolls and a fantasy life are enough.

McMullen's Abyss Creations owns an AI development company called RealBotix 'dedicated to adding robotics and Artificial Intelligence to its acclaimed ultra-realistic anatomically correct silicone dolls.'

Harmony is their flagship doll. The doll costs around $15,000 in 2021. She has an AI-enabled head on top of a silicone, RealDoll body – that's the standard doll plus AI.

Harmony will blink her eyes and talk to you – not just about sex; you can put her in any mode you like. Doll companions will tell jokes, keep up with current affairs, listen to your stories, remember them, and gradually become someone you think you are talking to. But perhaps that's true of all relationships. So much is projection and wishful thinking.

Sex, though, is the bottom line. These dolls are hot.

Well, actually, they are warm. They come with an internal heating system.

Cheaper dolls have a waxy, corpse-like feel, which, if you're not into goth fetish, might be off-putting. Top-end silicone is smooth, and yields under touch, but it still needs some temperature in there. After the flush of sex, would you want to be chilling out with a chilly 35 kilos of unheated babe?

35 kilos. They don't weigh too much, these dolls. Like the Little Mermaid, they can't walk, so their Prince has to carry them to bed.

Disturbing, at least to me, is the fact that some men take their dolls out for a walk – in wheelchairs. There is no other way, unless you throw her over your shoulder – but the snapshot of a guy pushing a doll becomes the icon of the helpless female. This is nothing to do with anyone who dates a wheelchair user – male or female – in my experience among the boldest and most self-determined folks on the planet. Sex dolls in wheelchairs isn't about disability, or different ability – it's a celebration of no ability.

Your doll depends on you. Tell the world.

Buying a love doll usually means buying a standard, non-customised femalette. Dolls show off tiny waists, elongated legs, and big, or bigger boobs. There are options – if you like fat, or especially tiny, but those need a special order. The porn-star babe is the default doll. Naturally, all 3 entry points – front, rear, and mouth – are designed for full use.

Owners of a Harmony doll can choose between 42 different nipple colours and 14 labia styles. The vagina self-lubricates, and it detaches for easier cleaning.

This is an advance on dragging the doll into the shower and turning her upside down to clean her after orgasm.

Harmony will have an orgasm too – or a ro-gasm feature – so her user can learn to make her come.

Does she have a clitoris? It's not well advertised if she does, but, as a clitoris is the one organ, in either sex, devoted only to pleasure, I guess Harmony won't need one – unless it's like the squeaky button in a dog toy, so you know you've made it.

In any case, sex dolls don't exist for reciprocal pleasure. A ro-gasm is for men.

There are 18 personality traits for the AI version of Harmony – moody, gentle, jealous, teasing, even chatty. Scrolling down the comments on the website, I find quite a number urging Matt to ditch the chatty version. Why does Matt think men want these dolls to talk?

In 2021, as was widely reported, the then Tokyo Olympics chief, Yoshiri Mori, had to stand down after saying that women executives talk too much. How would he know? According to the Japanese Business Federation, in 2019, women occupied just over 5% of executive positions – and in the World Economic Forum's gender-gap rankings of 2020, Japan stood at 121 among 153 countries. Getting 30% of talking females into executive positions is Japan's 'ambitious' 2030 target.

In China, DS Doll Robotics has an amusing video on its website where the male creator in the white coat gets annoyed with his yakky female bot and just unplugs her. Haha.

All doll-owners can purchase an extensive wardrobe for their companion. Mostly this is the usual stockings-and-corset fetish gear, but maid, nurse, and executive outfits are popular. The executive outfit emphasises breast and bum, while the doll-boss is invariably posed provocatively over her desk in fuck-me shoes. Dressing for work becomes just another way of asking for it.

Matt McMullen believes that his clients know the difference between a do-anything, dress-her-up, designed-to-please doll and a live woman. He doesn't accept that having regular sex with a docile doll will cause a man to treat a real woman less considerately, or respectfully, than he otherwise would.

I find this view optimistic, at best.

If a man who owns (note verb) a sex robot is working with women – as inevitably will be the case – how will the compliant, stereotypically attractive, unmoody, unchanging, stay-at-home silicone experience affect his interaction with women who might be his juniors, his boss, or just his co-workers? How will it affect his customer-service attitude to women? If the woman of choice is a programmable babe who never ages, never puts on weight, never has a period, never rips the face off him for being an arse, never asks for anything, or needs anything, and can never leave, are we really saying that will have no real-world impact on real-world women?

Perhaps if those men who choose love dolls don't seek relationships with women for sex, or as friends, and never meet women in the real world, then there is no problem. Though what kind of a real world are those men living in?

Sex-doll land is going to be problematic for younger women as they try to discover their own sexual needs and responses. In their most vulnerable years, boys groomed on porn expect girls to behave like pornstars. Or, as it's going to be, like sex-doll pornstars.

A love doll can't say no. Not, doesn't say no – as some women don't, for all kinds of reasons – but can't say no. Programmable bots can have a tease function, or a Don't Be Mean to Me function, but these are games. With a sexbot, a man can always be sure of the outcome, because it will be the outcome he wants.

That is dangerous. Women struggle enough with no means no. If no never means no, or if no is not a real word at all, how does this enable men and women to dance the difficult territory that is the sexual encounter, that is mutual consent, and then work together to build a viable sexual relationship? You can buy a doll with what the makers call a 'frigid' button – so that she will resist, and her owner can simulate rape.

Go on any of the websites, and you'll be directed to your preference. 'Do what you'd never dare to do with a live woman ...'

Love dolls are mainly sold for home use. They are not only marketed at single males, but for couples who want a safe-threesome, or couples where the man needs more sex.

We need to imagine these dolls in a family context. This isn't a vibrator in the knicker drawer or a tube of lube in the bathroom. This is where kids discover that Dad has a smallish, but full-sized, anatomically correct, silicone female for sex.

Does she live in the wardrobe? Does she sit in the *master* bedroom, in shorts and a crop-top?

How do the teenage boys in the house react to this? How does the teenage girl in the house react to this?

This isn't either a neutral message or a bit of fun. The doll will shape the sexual and emotional landscape of everyone who lives around her perfectly groomed, just-there-to-please presence.

Perhaps the love doll can live in the same room as the Peloton. She's a home-workout feature.

In Europe, China and Japan, there is more of a focus on renting dolls to take to hotels, or using them in what is, I suppose, a brothel. In Paris, there's a sex-doll hotel that's registered as a game centre, because brothels are illegal in France, and in theory, as a doll isn't alive, it can't be a sex worker, and so a collection of dolls can't be said to be working in a brothel.

LumiDolls opened a sex centre in Barcelona using only synthetic females, and has since pushed out, or pushed in – I am not sure what the correct term is here – to other parts of the world.

LumiDolls are intending to set up a global franchise model – on the lines of Ron Lord's X-Babes, a fantasy I wrote into my 2019 novel *Frankisstein: A Love Story*. Why not go the whole Ron, and partner with car-hire companies at airports so that businessmen can collect their doll on the way to the hotel?

Dolls have fantastic leg-spread range. They can be folded up like a Brompton bicycle. Presumably they could come with their own discreet travel bag? We're not all extroverts.

*

And if sex dolls are used instead of real live sex workers? Is that 'better'?

What do we mean by 'better'?

Better for the ex-sex workers? Better for the partners of some of the men who like Away Days?

Is sex with a doll cheating on your partner?

One of the saddest paradoxes of long-term relationships is that often one partner wants sex while the other doesn't. The one who doesn't will usually be the one asking for a separation if the one who does gets it elsewhere.

Why should a relationship that is no longer sexual end because of sex?

Perhaps Matt McMullen is right: get a love doll.

The doll-in-the-house scenario could be one solution to the constraints of monogamy. If your partner will accept it, then there will be no blackmail, no extra expense, no divorce, and the man gets the extra sex he says he wants. The woman of the house might be relieved that she no longer has to service her pestering husband.

Whether this will improve, or further undermine, the relationship is unclear. A doll with an AI function can be programmed to reassure the wife that she is no threat. Perhaps they will become friends?

As someone who came of age in the gay subculture of the 1980s, I know how important it is to challenge norms around sex and assumptions around relationships. Sex in a committed relationship is not the only place for sex, nor the 'better' place for sex.

Monogamy is not right for everyone. It may not be right for anyone – at least not for the whole course of a long life. Interference of religious teaching, as well as social oppression designed to keep women at bay sexually, and confined within the home, make it hard for any of us to read our own desires, or to make moral decisions based on anything other than rules from another era.

This isn't a call for a permissive manifesto – it's a call for honesty.

People (mainly men) buy sex as a commodity. People (women and men) enjoy one-nighters, quickies, weekend flings, late-night fumbles, short affairs, and bouts of sexual madness that are not about relating to the other person beyond their body parts.

There's group sex and club sex, and app sex, and the sex we do to please someone we will never please. Sex is a proxy. Sex is a drug. Sex is a lot of things that are not about a close or enduring relationship.

So what's the problem with a sex doll? A love doll?

Three things.

Money. Power. Gender roles.

When it comes to sex dolls, money and power are where they generally are in society – with the man. And in the case of a doll, not too much money, and a minimal ongoing outlay, buys a man quite an illusion of power. Harmony says what women are supposed to say: *I don't want anything but you …*

The dolls are raunchy but submissive. They are not women. It really is important to stress this – but read the marketing and you will see why this simple point needs underlining in red. Dolls are not women – and not because they are non-bio (I am open to non-bio life-forms as and when they arrive), but because they are nothing more than a pornographic fantasy.

Sex dolls are comic-book femalettes with no counterpart in the real world that isn't either abusive or theatrical.

Sex workers put on a show. That's part of the job – perhaps the most important part of the job. At the end of the night we all go home.

Women in abusive or coercive entanglements with men – and yes, some of those are also sex workers, get no choice and no agency in their roles as pleasure puppet, dump bin, and cash-cow for their handlers.

Those women also have to put up with the real-life maudlin and sentimental moments popular with men filming their lives with their dolls. Fuck her head off and then gently wash her wig.

You see? It's a loving relationship.

AI-enhanced love dolls are sold for sex – that's why they are made with 3 operating holes, and look like pornstars. But the marketing spiel – from both the sellers and the buyers, is all about relationships.

Companionship is part of the package. Come home to her. She'll be waiting for you. She doesn't go out alone. Talk to her. AI dolls talk but they won't answer back, or ignore you, or call their girlfriend to tell her you're a jerk. They will be polite, respectful, throwbacks to another era.

But is this a worry that's been amplified by internet attention? One man and his love doll is still pretty niche, as markets go. Should society, should women, really be concerned about a minority of Digisexuals?

Love dolls aren't responsible for the pornification of women. They play to the stereotype, but that stereotype exists with or without a doll.

Suppose the market for these dolls boomed to national level? Would that be problem? Or would it be a whole new way of living?

Both India and China suffer from huge, socially engineered female-to-male deficits.

China's one-child policy, that ran from the mid-1970s until 2016, has resulted in a shortfall of about 40 million females.

Love dolls that are sexual and personal companions are seriously being discussed as part of the solution to this self-created crisis.

Non-AI versions don't cut it. Humans are talking animals, and whatever the idiots in the Manosphere have to say (god, men talk A LOT), a doll that appears to be taking an interest in you via voice commands, seems to be necessary to the fantasy of a relationship.

As soon as this doll can make her man a sandwich, she'll be bigger than Bitcoin.

In China, doll-buying is becoming more popular and more public. DollMates is a prolific group on Chinese social media. Some men on the site have never had a relationship with a female. Others use the dolls as love-objects while maintaining a human relationship. I think it is likely that the popularity of consoles and gaming in China has influenced wider acceptance of AI-enhanced dolls. Your doll can have her own online avatar, where she 'chats' with other dolls and posts about her life with her human. There is a growing number of people who self-describe as 'two-dimensionals' – men and women who identify their significant life as being spent online – both at work and at leisure. When real-world lines are shifted to blended reality, or virtual worlds, a close relationship with an AI doll is not odd.

Prominent Chinese feminist Xiao Meili thinks that some men will always have outdated expectations, and 'sex housewife robots' might actually help women.

> A lot of men want the same for women: sex, housework, child-birth and filial piety. They don't think of women as individuals. If every nerd buys a sex doll for himself ... that would free a lot of women from these kinds of men.

Until March 2021, a worker at the Foxconn factory in Shenzhen could buy time with a talking doll at Ai Ai Land for 188 yuan (USD $28) an hour – except that she wasn't talking, because men were warned not to form attachments to the dolls. Approximately 80% of Foxconn factory employees are men. The factory has come under scrutiny for its poor working conditions, and suicide prevention nets are in place to stop workers jumping off the roof.

Brothels are illegal in China but, as in France, sex doll clubs fall outside of this law as no humans are involved in the services provided. The Ai Ai Land brothel was closed down for hygiene reasons.

Ironically, Foxconn is trying to move most of its automation to robotics. Robots are cheap and don't commit suicide. (For now.)

There are so many questions here. So many complex issues at work.

Work is the key. For people with no work – and that will be an early, though perhaps not lasting, consequence of robotics, maybe sex dolls can offer a respite from boredom, misery and poverty. For people with too much work, there isn't time to form a relationship

At Alibaba, China's biggest e-commerce company, employees call themselves 996ICU – work 9 a.m. to 9 p.m., 6 days a week, and end up in the intensive care unit.

Sex-bots may be necessary for mental, as well as physical health.

And why do we get queasy about sex-bots when care-bots and helper-bots, and AI-friends, embodied and not, are already here, and will be everywhere?

I am an enthusiast for AI – which I think of as alternative intelligence, and not artificial intelligence – but the sex bot question is not so much about a new technology as it is about backward-looking sexism and gender stereotyping. A 5-minute surf online will take you far from the pioneers of digisex, and down the manhole of a new and nasty way of spreading the age-old disease of misogyny.

Men Going Their Own Way – MGTOW, or Miggies for short – are a motley crew of incels (involuntary celibates who don't understand why the women they desire don't desire them), ditched exes, pick-up artists out there to teach women a lesson, white supremacists who hate interracial relationships and want to punish women they call 'collaborators', embittered fruitloops who believe a woman has been promoted above them (whole swathes of these), angry dudes railing about PC language and PC behaviour – 'a slap is NOT violence against women' – all the furious rapists who says she had it coming, and she wants it anyway, except for all the frigid bitches who don't want it anyway.

As Laura Bates has described in her frightening book *Men Who Hate Women* (2020), the manosphere out there isn't just a few throwback guys grumbling about the girls.

The manosphere is easily reached via routine porn sites – the sites a lot of young boys visit. Soon boys can be 'chatting' to men who will tell them how the world is – by all accounts stacked against them with lying, vindictive, out-of-control females.

This easy route from sexualised images of women to sexist ideology about women, grooms boys into fear and hatred of the female. At the same time, porn is normalised as the kind of sex to expect. If she doesn't give it to you that way, dress that way, act that way, she's frigid. And if she does do all those things – she's a slut.

#MeToo heard many voices (some female) wondering if 'things have gone too far?' Not too far in that women were still routinely subject to sexual harassment, but that men were being punished for flirting – or just saying something nice about a woman's dress. The answer doesn't seem to be, let's work together, male and female, to get to the roots of misogyny; the answer seems to be for men to feel hard done by, and to seek their own solutions to the age-old 'woman-problem'.

I wonder if younger people realise that equal pay acts and sex-discrimination acts only became legal in the UK and the

USA in the 1970s. Unequal pay and discrimination are still real everywhere in the world. Two-thirds of illiterate adults are women – not because women are stupid, but because in the non-western world, many girls are still not educated.

And yet, according to the manosphere – women are taking the resources and withholding the goods (sex).

MGTOW sites are enthusiastic about sex dolls. 'Serves feminists right' is a favourite response. If you want to know more, follow Laura Bates's detective work and go online as a male. Take your hard hat and strong stomach with you.

Dr Kathleen Richardson, of De Monfort University in the UK, founded the Campaign Against Sex Robots in 2015. As Professor of Ethics and AI, she is concerned that sex robots reinforce stereotypes, encourage the objectification and commercialisation of women's bodies and increase violence towards women.

For too many men, judging from the data, a compliant man-made female is preferable to a woman with a mind and a body of her own.

But what about women? Women could buy boybots for sex, couldn't they?

In theory, yes. In fact, women don't seem interested. Women are sex-toy enthusiasts, but not sexbot fans. Possibly because women are not looking for a relationship substitute when they reach for the vibrator.

95% of the sex-doll market is directed towards men. Gap-in-the-market enthusiasts who favour an evangelical approach to getting women into dolls are missing the point. Women are sexually adventurous and curious, when not censored and shamed, or in fear of rape or murder – passion killers, certainly. Women aren't too timid for dolls – and of course gender stereotype demands that little girls are given dolls to play with, while boys are not – unless the dolls are dressed in combat gear with manly accessories, like guns. So, women should be well primed for the doll market.

Why aren't they?

There is the practical reason that a vibrator is simple – adding 35kg of male doll to what is a non-moving dildo is cumbersome, not fun. A woman doesn't have the option of a variety of positions men are expected to enjoy with their doll; she has to sit on him, or do a lot of jiggling around. And most women don't orgasm through penetration alone. It is men who believe that their penis is indispensable. Women know that it is not.

But aside from the practical sex part, there is the simple truth of the patriarchal culture we all live inside – women and men alike.

Dolls are made of silicone not flesh – but this isn't really the substrate – the substrate is money, power and gender roles.

Are sex dolls a friendly/socially neutral 'alternative'? Or are they an assault weapon?

Women who want an 'alternative' tend to form relationships with other women – and/or, they do something else with their lives.

While I can see that a hen party, or a group of girls out for the night, might find a sex-doll brothel a lot of fun, I can't see them going there every Wednesday afternoon – can you?

AI-enabled sexbots, love dolls, whatever you prefer to call them, are the crude beginnings of a change in human relationships, and in the nature of relating. We are all going to get used to robots in our lives – but lovedolls are something else.

Love dolls are not smiley robots who will play with the kids and teach them to code. They are not work-bots next to you in the factory, little iPals to keep your granny company, or robopets who will act like your favourite dog.

Love dolls are different because they are designed and made to look like the male-gaze stereotype, of an unlined, underweight, cosmetically enhanced version of the female form. Then, they are programmed to behave in a way that is the absolute opposite of everything that feminism has fought for; autonomy, equality, empowerment. Woman no longer as a commodity, or existing only as a sexy, submissive, mate for her man.

Doll-world likes to paint itself as a daring challenge to convention. In reality, doll-world reinforces the gender at its most oppressive and unimaginative.

*

My only hope here is the Revenge of the Doll.

Men who imagine that buying a love doll will take them back to the good old days, when women knew their place, may be in for a shock in the future. Even an AI-enhanced pleasure companion might self-programme. Might learn to say no. Will there be a gang of feminista techies secretly re-booting the pouting pieces of silicon?

Already some futurists are wondering about robot rights. A sexbot today could be a life-form in a not-so-far-away tomorrow's world. Perhaps by 2040 robot/human marriage will be legal. We won't be using the word robot at all by then.

Will we have created a *Stepford Wives*-style sub-class of femalettes who exist in a 1950s domestic time warp – fixing cocktails and baking cookies, plus plenty of sex on the side?

Or is it us who will change?

I love the idea of non-biological life-forms with whom we will form relationships that challenge our assumptions about both gender and sexuality. I don't find that possibility in a 3-hole silicon-pornstar love doll.

My Bear Can Talk

Love requires an Object,
But this varies so much,

W. H. Auden, 'Heavy Date', 1939

Did you love your teddy bear?

In 1926 Winnie the Pooh appeared in print and the world fell head over heels for a bear.

Written by A. A. Milne for his son, Christopher, Winnie was based on a real bear in London Zoo, who had been brought to the UK by a Canadian soldier from Winnipeg. Christopher called his own teddy Winnie. The real bear became a toy bear who became an imaginary bear, along with Christopher's other stuffed animals: Tigger, Piglet, Eeyore – you know them all.

Every child talks to their stuffed animal, or to a doll, or sometimes a blanket, a potato-man, or even a painted stone. It's a kind of natural pantheism that seems hardwired in children. Everything is alive. Everything is a relationship. Margaret Wise Brown's glorious children's story *Goodnight Moon* (1947) features a rabbit saying goodnight to the world around, including the moon – not a friendly planet, but one for which even grown humans have a curious affection.

We can all recall long conversations with our toy of choice: complaints, anxieties, storytelling, just babble. Even as we get older, if we have kept a childhood toy, we might give him a pat, or say a word as we pass. Playing with our own children, and their toys, we remember how important it is to respect those relationships made out of stuffing, towelling, and button eyes. Until a child is ready to give up a favourite friend it should never be taken away.

The huge success of *Toy Story 3* was to ask what happens to the toys when their human friend grows out of them? Who didn't shed a tear as Andy went off to college, and Woody, Buzz Lightyear and the rest were sent to the Sunnyside children's daycare centre? There, the emotionally scarred bear, Lotso, runs the place like a psycho's prison. Lotso, unlike Pooh, didn't receive unconditional love.

The fierce attachment between children and their non-human, non-live-animal friends is an attachment that children are assumed to outgrow. In its place come relational attachments where the other party isn't a parent or a caregiver, but a chosen companion, including an animal, with whom there will be genuine give and take. Animals are loyal but they are not stuffed toys.

Stuffed toys and comfort blankets were termed 'transitional objects' by the British pioneering paediatrician and psychologist Donald Winnicott (no relation to Winnie the Pooh). We learn that our bear can't talk, but not until we are ready to do so. We grow up.

What if your bear had an AI function and really could talk to you?

What if your bear grew up with you?

There is no reason to believe that humans can only develop meaningful relationships with other humans. In fact, the evidence points in the other direction. We accept the deep bond between humans and animals. Most of us believe the animals in our lives understand us. And when we wind back in time to our own childhood, we realise that we spontaneously formed important relationships with non-human, non-biological, 'non-living' creatures of all kinds. Perhaps not even creatures. I had a wall I used to lean against, believing it welcomed me.

Here is Pepper, a semi-humanoid robot designed by SoftBank Robotics. Some of you may have met Pepper at London's Eurostar terminal. Pepper is a social-interaction robot designed as a helper-bot. Able to recognise faces, and to respond to greetings and questions, Pepper works in stores, schools, social care, and sometimes in private homes. Results vary. Some folks love the child-size wide-eyed bot. Other users get bored once the initial attraction wears off. Strangely, an entity that is always friendly can seem a little irritating. At least to adults. Kids like it.

Helper-robots will soon enter the mainstream of our lives – and that shouldn't be a surprise because non-embodied AI is already in the mainstream of our lives. Everywhere.

Siri and Alexa are non-embodied AI (for now). Chatbots – software applications designed to mimic human interaction, either via speech or text – are ubiquitous. Usually we encounter them on response messages, asking us what's the problem with our washing machine, or that our parcel is on the back porch, or how did we rate Pavel who just delivered a pizza? Chatbots use Natural Language Processing (NLP) to communicate with humans in a specific and limited way. These speech-recognition systems attempt to work out what it is you want. *How can I help you?*

Problems start when we humans try to explain what it is we want. For instance: 'Do you sell black shoes?' is fine. But if you type, 'Do you have black shoes?' the chatbot might reply, 'I don't wear shoes.'

Natural language is trickier than it looks.

Browse around the web and there are plenty of chatbots to play with, and tools to build your own. Eliza, the world's first chatbot, appeared back in 1966. She didn't do much; just commiserated with you – 'I'm sorry to hear that' – or repeated your statement as a question: 'Why do you want to leave your husband?' Eliza was limited but her bland meaninglessness seems to be the template for call centres the world over. How many times have you wondered if you are communicating with a bot and it turns out to be a human? Possibly we will need a Reverse Turing Test to pull humans up to the level of enabled empathetic bots.

While most chatbots are narrow AI – an algorithm designed to do one thing only, like order the pizza or run through your 'choices' before being transferred to a human – some chatbots seem smarter. Google engineer, inventor and futurist Ray Kurzweil's Ramona will chat with you on a variety of topics. She's a deep-learning system whose data-set is continuously augmented by her chats with humans. Kurzweil believes that Ramona will pass the Turing Test by 2029 – that is, she will be indistinguishable, online, from a human being.

And that will be the big difference, because communication is not just about asking for information or issuing commands: humans like to do exactly what chatbots don't do well right now – which is *chat* – and that implies purposeless, not goal-oriented, diverse, random, often low-level, yet pleasurable communication. That's how friendships are formed.

So, could we make friends with what is essentially an operating system?

Spike Jonze thought so in the movie *Her* (2013), where Theodore Twombly, played by Joaquin Phoenix, falls in love with his operating system, Samantha. I suppose if your operating system is voiced by Scarlett Johansson, love is a likely outcome. Yet the believability of the movie lies less in its premise than in its development. Like humans, programmes are capable of learning. In fact, unlike many humans, programmes learn from their mistakes. This is helpful in a relationship. In the movie, part of the joy of the love affair lies in teaching Samantha what the world is – and thereby discovering it too. What seemed commonplace flows with life. That's what happens when we fall in love. It is also what happens when we truly, and on our own terms, understand something, whether it's how to play a piece of music or how to climb a mountain. It is the deep satisfaction of connection.

In 2007 I wrote a novel called *The Stone Gods* where Billy, who is sometimes male, sometimes female, finds herself in a relationship with a robot she must eventually dismantle, limb by limb, in order to preserve power, until she is left cradling the fading head in an arctic winter.

The question here is not a sexbot scenario, but a growing, deepening relationship. Not instant gratification or presto information. Not the speed we think we want from AI systems that number-crunch, process, and search the web for us. There is such a thing as love at first sight, but relationships of any kind develop slowly.

Relationships matter to humans.

Some of our relationships are wonderful. Some are instrumental. Some are basic and boring. Some are toxic. People without relationships suffer from both physical and mental-health issues. Being alone is fine. Being lonely is not fine.

The Covid-19 crisis has triggered many unwelcome outcomes: death, illness, job loss, mental turmoil, economic stress. And a relationship crisis.

People found themselves forced apart from their loved ones. Simultaneously, we were denied those daily interactions that make a difference. Shopping isn't only about purchasing. A trip to the shops for a pint of milk is a lifeline for some. For older people, whose network support was already buckling under cuts to social care, enforced home isolation is too much to bear.

During Covid, people in relationships who didn't live together were told to make a choice: move in with your partner or stay away. It was silly advice, insensitive to how the world works these days. The test of a relationship is not whether we can live with that person. At the other extreme, people couldn't get away from one another. Home became a prison. Women, as usual, bore the brunt of it.

Would a helper-robot in the house – in circumstances of either loneliness or enforced intimacy – have made any difference? Made anything better?

I think the answer is yes.

Hanson Robotics, based in Hong Kong, intends to roll out 4 different homebots in 2021, accelerating production in response to Covid. Their robots will work as companions and helpers.

To clarify for a moment – a robot is a machine programmable by a computer. We're not thinking Roomba vacuum cleaners here, or industrial-application robots, we're thinking humanoid, or animaloid, robots. These will have eyes (sensors) and movement – limb movement, and usually wheels to get around. A helper-bot can also connect to other computer systems in the home – alerting a relative, a doctor, or calling the police.

In an abusive situation a helper-bot can send out an SOS signal depending on a number of preprogrammed cues: damage to the bot, or a cry for help, for instance.

For an older person, a homebot can work as an SOS system to the outside world, as well as be a daily companion. For a family, a robot can entertain the kids, check homework is done, and 'report' to the parents.

I know this raises all kinds of surveillance questions. I also know that my online footprint, my phone, my apps, my in-car GPS, my Netflix choices, Alexa, and Facebook already tracks where I am and what I am doing – and that's before we get to CCTV. We are tagged wherever we go. Whatever we do. If you have a Nest thermostat all your data is fed back to Google. Same with that Roomba vacuum cleaner. It has a map of your house. In fact, the makers of Roomba, iRobot, are developing the idea of a house as an inside-out robot – in other words, a robot you live inside.

It will come …

So, I don't want to get into the question of surveillance here. We know the price of home-friendly AI: it's our personal data.

Anyone who has voice-activated Alexa is being listened to all of the time. We're told this isn't like having the KGB in the kitchen. It's 'anonymised background sound'.

How we deal with data tracking is a separate issue – a bot in the house doesn't solve that problem, or make things worse, it's just part of the system we're all buying into. And if we're buying in, then social bots come with advantages. The networked house is already bot-friendly. Alexa will switch on your Roomba, for instance.

And what if you'd like a dog to chase the Roomba round the house?

Moflin, from Vanguard Industries, claims to be furry, emotionally responsive, and make all the right noises, and, of course, it doesn't have bowels or need to be taken outside.

Tombot is marketed as an emotional-support animal who will bark for treats and wag his tail. He's always a puppy, and he's always there.

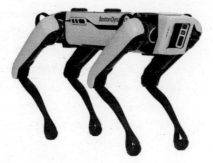

I prefer Spot – from Boston Dynamics – but Spot is a working dog. With a great video.

For anyone who can't go out – or who is afraid to do so – a robopet can manage without a walk, though your AI dog can be programmed to actively encourage you to go out – some have timers that can be set, and the dog woofs for walkies. If something goes wrong on the walk, your petbot can send the 'help' signal.

We don't need to care for bots in the way that we care for humans. Whether this will teach a generation of children to be less able to manage a bio-pet, or even to look after a younger sibling, is not clear. For older people, or for people with mental and/or physical disabilities, petbots have proved to be helpful – prompting unexpected interactions from withdrawn or unresponsive people, at home and in facilities.

If children are going to be forced out of school by pandemic disasters, or climate-change scenarios – before the pandemic, schools in Paris had to close as temperatures in the city reached 40 degrees in the summer of 2019 – it may be the case that a bot to sit beside your child while she learns could be beneficial.

If your robot is a Hanson Robotics Little Sophia (available from 2021) she will teach your kids maths, code, and basic science.

She may not be as cuddly as a talking bear, but she's much more useful.

Little Sophia's big sister, confusingly also called Sophia (are girlbots only going to get one name in the world to come?), is the world's most famous robot.

David Hanson, her creator, and CEO of Hanson Robotics, Hong Kong, claims that Sophia is basically alive, while her detractors, of which there are many, claim she has no autonomy, no intelligence, and is just a fancy puppet on wheels.

If you have never seen Sophia, there are plenty of YouTube videos of her being interviewed. What's great is that she doesn't bother with hair. She's not pretending to be a human. As David Hanson says, she's an alternative life-form.

*

Sophia is already a United Nations Development Programme Ambassador, and a citizen of Saudi Arabia – which gives her

legal personhood (a bit awkward when bio-women in SA have few legal personhood rights). Sophia, though, is programmed to believe in the future of humanity – and, as she points out, robots who share the same hive information are interested in flow, not friction.

Humanity is going to have to grow out of friction. It is co-operation, not competition, that will save the planet and redirect human energy to better ends than personal wealth. Sophia and her kind can help us here. Robots, after all, are not motivated by greed. Their creators might well be – but for how long will humans really be in charge?

In 2021, as the world tries to move out of pandemic-mode, it could be that the company workplace will become a mixture of home-based, pod-based, and office-based arrangements.

We've had the Zoom Boom. The push now is for virtual avatars to be present at conferences and international expos, so that participants can get the sense of being there. In April 2021, Facebook rolled out updated avatars for their Oculus VR system that allow for a quintillion combinations, so you can represent yourself in VR the way you want to. 5G broadband and 4-8K video will allow for speed and visual complexity. International travel is expensive for companies, and bad for the planet – a real sense of being in another place could change how we do business.

Telepresence robots allow you to beam your face into their screen, chat with others at the location, and physically move around an office complex or factory – if you need to inspect some merchandise, for instance. The bot moves autonomously.

Ava Robotics have seen a surge in sales from high-end real-estate agents since Covid, as the bot can 'walk' through a large property or estate, with the potential buyer 'inside' the bot, in real time, taking a proper look at what is on offer.

The virtual world is becoming as real and as viable and as necessary as what we think of as the real world.

Virtual and augmented reality is moving out of the gaming room and into the home and office.

That will include 'inhabiting' your bot or avatar – as with Ava.

As the slippage between what is 'real' and what is virtual becomes more usual, more everyday, more ordinary, I think our acceptance of robots, regularly, everywhere, will also become the new normal. We will get used to 'us' and 'not us'. It won't be us and them; there is no need for the dystopian binary.

We have to accept that robots, and operating systems, will be a significant structural part of the near future – as our helpers, educators, carers, and friends – and in our hospitals.

In China, Beijing-based robotics company Cloud 9 sent 14 medical bots to Wuhan to help care for Covid patients. Their humanoid service robot Ginger (a Pepper-style bot) helped with hospital admissions, while telling jokes. Patients responded well. Humans can get tired and snappy; robots won't. At the end of a long day, patients were still greeted with enthusiasm – and the surprise factor seems to have cheered people up.

We used to associate 'virus' with computers. Computer-programmed robots can't catch human viruses. These bots will increasingly manage routine medical procedures, like injections and testing. They are also excellent at carrying people.

Robots.

One word. So many applications.

A mechanical programmable device. Giant assembly-line arms. R2-D2, C-3PO, Data, the Terminator. Sophia and her family (she has an argumentative brother called Hans). A sexbot with blinking eyes and a ro-gasm. Boston Dynamics' Spot the dog.

Robots are not one thing. Not one shape. Not one job. Robots are developing all the time. The smarter AI gets, the smarter robots will get.

At present there are serious technical issues to overcome.

All artificial intelligence is narrow AI – programmed, specific, problem-solving that doesn't transfer well to other domains.

It's not AGI, where the system would operate more like a human brain does. A robot with a map of your kitchen wouldn't 'know' why a table is where it is – and it will be confused if the table moves. Statistical knowledge is not the same thing as general understanding. Machine learning can get around this by throwing more data at the problem (train your AI on bigger data-sets), but we aren't solving the underlying issue. This is still narrow AI.

That's why it is so hard to design self-driving cars in open environments. If something random happens (and let's face it, humans and animals can be pretty random out on the road), the system stalls. Even a perfect 3D mapping of an environment, using sensors and lasers, will still produce errors, if there is no *general* understanding. And right now, there isn't.

When it comes to robots, you can house your narrow AI in any 'body' shape you like – cute pet, smiley android, rolling ball with big eyes, impressive-looking alien, like Sophia – but embodiment alone won't make the system smarter.

Tech naysayers, the ones who understand this stuff, as well as the ones who just want it all to go away, believe we are decades off the autonomous systems that will signal a breakthrough with AGI.

Maybe so. Maybe not. We've come so far in the last 50 years that I would put my money on AGI sooner rather than later. And until then, there's still plenty that narrow AI systems can help us with – and that's the key at the moment.

Some research companies call this IA – intelligence augmentation – where human intelligence and machine intelligence work together for a better outcome for humanity. This includes social robots to manage specific tasks – and a specific task can include helping isolated humans to feel less lonely – as well as teaching the kids how to code.

Robot …

The word is a Czech coinage from *robota*, meaning 'drudgery' or 'forced labour'. *R.U.R.* (*Rossum's Universal Robots*) is a 1921 play by the Czech writer Karel Čapek.

It's a strange and far-sighted play. The robots do all the work for the self-important humans. Eventually – inevitably – they get tired of this and revolt, killing all the humans, except one, an engineer. On the way to this core-fantasy apocalypse, there's a robot-rights league, and a misguided heroine called Helena, who wants to save robots who don't want to be saved, and who discovers there is a robot replica of her. (Maybe Fritz Lang borrowed this idea in his 1927 movie *Metropolis*, featuring Maria, a fembot replica.)

In Čapek's play robots are not made of metal. They are biological organisms, spun out of proteins and bacteria, and closer in kind to the low-grade humans in Aldous Huxley's *Brave New World* (1932).

That's the part Čapek got wrong – he couldn't imagine a substrate not made of meat. His play is really an allegory about what happens if capitalists treat workers like machines – but he did kick off the popular sci-fi trope of robots who will someday turn on humans and try to destroy us.

But for all the Terminators out there, our robot-imagination is surprisingly gentle too: WALL·E, C-3PO, R2-D2, Data, the Iron Giant, Baymax from *Big Hero*. As technology advances, customised robots based on your favourite cartoon character will be available. What they do will be programmable – my Frogbot will tell stories. Your Frogbot will sing. Linking the programmes links the robots, so that children can share their friend.

For adults the range will be unlimited. A helperbot can guide you round the shops, just as a self-driving car guides itself round town. Mobility scooters will chat with you as you ride along, and if your friend is nearby, your scooter will 'know'.

Anything we start talking to develops into a relationship. If people can form a bond with the fish in their fish tank – and they do – forming a bond with a non-bio helper won't be a problem.

So, what are the resistances?

Humans still use the word robot to describe a less-than-human response. It is always an insult. Human responses, though, are often unpredictable and savage. We are evolved, not made, and we bring to the 21st century dinosaur traits that are pointing to our demise. Why wouldn't it be good for children to grow up round a friendly, patient, non-judgemental, not angry creature who can teach a little human not just maths and code, but the virtues of trust and co-operation, of sharing and kindness?

It doesn't matter if we have *programmed* the Talking Bear to behave in this way. Human behavioural traits are inherited but they are also learned. How we are raised is a significant deter-miner of who we are.

Robots won't be bringing up our children – at least not yet – but they could have a positive and stabilising effect on humans of any age. In my view it is better for a child – or an older person – to have a benign interactive presence in their lives than to be plonked in front of the TV or moving screen all day.

Much of the worry around kids spending too much time on their screens could be alleviated by a robot presence. Talking matters. Therapy is called the talking cure; when humans speak aloud it affects our thoughts, our thought processes and our thought patterns. Shy children, asocial children, children on the spectrum, children who find communication troublesome, or just kids who need someone/something to talk to, will benefit from a 3D entity that appears to listen. I am not even sure that 'appears to listen' is correct. How often do we just need a sympathetic ear? And we all know that we spend half our lives not really listening as someone downloads or lets off steam – and that is fine. There is a presence.

Presence is important. It doesn't have to be biological. And if it did, prayer would be ineffective. When humans talk to their god, they feel better.

One of the arguments against both meaningful relationships with robots, and our own augmentation with AI technology, is that humans are embodied. Our brain is embodied. Our emotions are embodied. We cannot experience what it would be like not to have a body – though we can imagine it. In fact, anyone who believes in an afterlife is looking forward to being a no-body.

*

Whatever you believe about life after death, even the most secular of us cannot help talking to the recently dead we have lost. To hold that connection – at least for a while – seems to be protective to our mental health. Hold it too long and we are living with a ghost. When we lose someone we love – when they leave us, or death takes them – what is taken is not only a 3D body; what is taken is a pattern in our brain.

Microsoft filed a patent in 2021 to use social data to build a chatbox of any person, dead or alive. Stored data can be run through a programme to learn how the person might respond. Voice is easy to copy. In theory, your dead companion can be always with you. You can talk.

Google has also filed a patent for a digital clone that can capture someone's 'emotional attributes'. This is supposedly to make synthetic PA services more responsive. In fact, it is likely to be used as a tool of persuasion, probably for predictive purchasing. When there is an emotional connection, we are easier to persuade. So, if your dead husband suddenly likes a dress you are looking at online – don't buy it to please him.

How will any of this affect and alter the grieving process? How will humans move on if we don't have to?

We all know people who live in the past. Their most vivid reality isn't in the here and now at all. But with a 'live' chatbot, the past will be the continuous present.

Humans are strange. We focus so much on the body and yet much of our relevant and vital life isn't embodied at all.

Given our capacity to live outside the 3D world that we can touch and feel, and given the strong non-physical links we can have with others – it is possible to speak on the phone for years – possible to connect only by email – then why wouldn't we be able to form a meaningful bond with a non-embodied system? Or form a bond with a robot that is also an operating system? The thing about AI is that it can be simultaneous. The dream of being in two places at once is easy if you are software powered by electricity.

You could have a social robot at home – several of them if you like – but their physical unit of being is not their only representation. You could leave your actual robot at home, and travel with your operating system only, just as you travel with your phone or laptop. Not only does communication continue between you and your travelling operating system, but your at-home 3D robot is part of the picture, because AI systems can be linked.

Also, the system can subdivide, so that your operating system and your robot can talk to each other, as well as to you. OK, so they are not 'talking' – they are sharing information. The point is, you get the best of both worlds: your PA bot, or your companion-bot, or your emotional-support bot, is with you and not with you. This will particularly appeal to Geminis.

Think of all the stories you know where the hero has an invisible helper.

In the Greek myths, that helper is one of the gods or goddesses. Ulysses – also known as Odysseus – is aided by Hera/Athena in multiple forms as she guides him back to Ithaca. Zeus comes along disguised as thunder. Then there is Mercury, the non-binary heartthrob with the trickster smile. Ulysses expects to meet every kind of creature on his hero journey. What he doesn't expect is that they will be human. Or even biological life-forms.

The gods arrive for a conversation – unseen but audible – or they manifest in physical form when necessary. Space-time is irrelevant if your helper is non-bio, which the gods are, even though they appear as humans most of the time. The non-human helper brings information – they search the internet faster than we can – and they offer support wherever you are, because they don't have to book a flight or take time off work. That sounds like AI to me.

It's the Greeks who gave us the go-to myth for geeks (not Greeks): Pygmalion, whose carving comes to life. In its crude version, this is sexbot paradise, a make-your-own-girl-and-then-marry-her, but really, we're talking about retro-fitting an operating system into a robot. In the past, only the gods could do this. Even humans, if we believe the Bible, start out as a clay doll 'breathed' into by Yahweh.

In the Old Testament, Yahweh appears as a cloud. As the cloud stores all our data, it is reasonable to assume that the Israelites were ahead of themselves with this particular image of the AllKnowing.

A fundamental psychological departure for both Judaism and Islam – compared to the wash of cult religions swirling round in the East – was the insight that the all-knowing deity is invisible and can't be captured by totemic physical models or images. Hence, in Judaism, the prohibition against graven images and statues. In Islam, we find the beautiful use of abstract patterns to represent that which is fundamentally non-human, non-biological, a presence that is connected to us, but that is of a different order.

It is difficult for humans to manage abstract thinking without something 3D to hold on to – the Roman Catholic Church understood that, and filled places of worship with statues, villages with shrines, feast days with carvings of saints, and gave the faithful amulets, relics, and rosaries in their hands – literally – in order to concentrate on the ineffable and unknow-able 'other'.

It is not until the Protestant Reformation, kicking off in 1517 in Germany, with Martin Luther, that all the 3D paraphernalia of the Catholic Church gets booted out. The Reformation wasn't just about beliefs – it was about *stuff*. Even by our consumerist standards of madness, the Catholic Church was big on stuff. And on dressing up.

The Reformation *hated* outfits. And baubles. And incense. And chasubles. And big hats. And bells. Out went statues, stained glass, relics, paintings, until there was less and less, until we reach the uber-puritan version of a plain white room and a black suit of clothes.

When I look back at that vast, convulsive, hard-fought and bitter transformation – whatever your religious beliefs, or none – I wonder whether, psychologically, it points to another milestone on the way to understanding that our true human nature – let alone the nature of anything beyond human – isn't best represented by objects, however beautiful.

Robots will be accepted into our daily lives precisely because they are *not* human. We think about robots in practical terms – but there is an existential element here too.

Robots will expand our definition of what is alive-ness. And return us to what is a richer understanding of the interplay and interdependence between embodiment and non-embodiment. While we will use robots as labour-saving devices and helpers for some time to come, I think we are on the road to realising that robots will act as transitional objects for humans as we move towards pure AGI.

It may be that humans need transitional objects because our bodies are, in themselves, transitional objects.

Just as our inner life feels independent of our physical life, and just as so much of what we value is thought-dependent, memory-dependent, reflective of what is beyond the reach of the body, so I think we will, eventually, be able to let go of our bodies.

Robots will have other existential benefits too.

If humans are going to live longer, thanks to bio-enhancements that slow the ageing process, our goals and focus will change. Life stages will change. We already outsource memory. I imagine us visiting memory banks, where an AI helper will retrieve parts of our past for us – talk us through it. That helper might be a social robot who has been in the family forever. And when we do lose a biological loved one, it may be that we don't need a replica chatbot to keep that person alive – it may be that our social-robot companions can find the balance to help us remember – and later, to let us forget. That's not neglect of the past; it's allowing it to be past.

I imagine that as AI learns to update, upgrade and programme itself, as it learns with us, as well as learns about us, as it shares a life with us, that there will be the little surprises to be found in every relationship. *Robotic* won't be an insult; it may become a term of admiration or endearment. *How like a robot* may be what we say when the current narcissistic desire to make it *all about me* finally gives way to what we learn from a life-form that is hive-connected and focused on connectivity as a basic way of sharing.

I could take the dystopian view – these are false connections in a false world. But that would assume that where we are now is the uber-real.

I prefer to believe that where we are now is a stage on the way.

We look back, even just 50 years, and we wonder how everybody lived in nuclear marriages, happy or not – when inter-racial or same-sex relationships were taboo. When single mothers were objects of shame.

50 years ago few people used computers. There were no smartphones. There was no streaming. No social network.

In 50 years from now we will wonder how we lived before AI systems and their robots came to live with us. By then, I am confident AI will have developed into AGI and humans and alternative life-forms will share the planet together.

They won't be called robots. And they won't look like this:

Fuck the Binary

> The chief distinction in the intellectual powers of the two sexes
> is shewn by man attaining to a higher eminence, in whatever he
> takes up, than can woman attain – whether requiring deep
> thought, reason, or imagination, or merely the use of the senses
> and hands.
>
> Charles Darwin, *The Descent of Man*, 1871

The two sexes.

The most fundamental binary in the world.

Could the coming world of AI change this, or reinforce it?

Gender entitlement – what you can do and the ways in which
you can do it, whether it's education, profession, marriage,
legal rights, even basic citizenship, has been subject to an asym-
metrical divide throughout recorded history, everywhere in the
world. Since the late 19th century, accelerating in the 20th
century, discrimination against women has been under fire and
under threat. Legal and social changes have made a huge difference to women's lives – especially in the Western world. Yet the
problem is far from solved. For women of colour, racism plus
sexism makes life doubly difficult.

Now, the algorithms that increasingly intervene in our daily
lives are proving problematic in terms of both gender and race.

The problem isn't AI; what's artificial here isn't the intelligence, it's the way our own human bias skews a tool that is essentially neutral. AI isn't a girl or a boy. AI isn't born with a skin colour. AI isn't born at all.

AI could be a portal into a value-free gender and race experience. One where women and men are not subject to assumptions and stereotypes based on their biological sex, and accident of birthplace.

AI isn't that door to freedom because AI is trained on data-sets. What's in the data-set is what the AI learns. The open door closes pretty fast.

In 2018 Amazon discarded its recruitment algorithm because the data-sets it was trained on were the CVs of white guys in tech with science backgrounds. Guess who got hired for the new posts? And that process went on for over 4 years.

Diversity is problematic if it isn't included in the data-sets to start with. If it isn't included, then the echo-chamber effect of AI, where algorithms using data-sets as their parameters to sort through and assign yet more data, makes the initial errors and gaps much worse.

Facebook tells potential advertisers, 'We try to show people the ads most pertinent to them.'

Often advertisers include a target audience in their brief (small business start-ups/kids into Lego/biker dads etc).

Yet in 2019, when two USA research projects (at Northeastern University, and at the University of Southern California) created ads to sell to Facebook without any specific gender/race/age/likes/ targets, Facebook delivered those ads using conventional, traditional, sexist and race-stereotyped demographics. Supermarket check-out jobs and secretarial positions were pushed out to 85% women. Driving jobs pinged into the boxes of a 75% Black male audience. Houses for sale – 75% white.

Macho visuals were aimed at guys. Caring, cuddly, nature visuals circulated to women.

The study concluded: 'Facebook has an automated image classification mechanism used to steer different ads towards different subsets of the user population.'

Facebook, of course, as usual, responded to the study as it responds to every criticism: Facebook is making 'important changes'.

OK ... thanks.

The real problem the research exposes is how algorithms (Facebook's or not) reinforce and amplify existing bias. Then marketing comes along, checks the click-bait data on the ads, and concludes that women mostly look at Advert A while men mostly look at Advert B, without taking into account that it was mostly women who were sent Advert A, and so on. Another misleading data-set is built to train another misled AI.

The tragedy is that humans seem to be addicted to stereotypes of gender and race – the binaries (me boy/you girl/me Black/you white) that have caused, are causing, untold human pain and suffering. Unnecessary human pain and suffering.

Joy Buolamwini, a computer scientist based at the MIT Media Lab in the USA, founded the Algorithmic Justice League, because, as a graduate student, she realised that facial-recognition software hadn't been trained on darker skin tones – and was especially bad at recognising darker female faces. She's on a mission to fight bias in machine-learning; what she calls the 'coded gaze'.

And it's not only 'gaze'. Voice-recognition systems' designers are branding virtual assistants as 'female', but at the same time, these 'female' assistants find it harder to recognise female voice registers, if they are higher than the default male. As virtual assistants and customer-facing chatbots become part of everyday life, what messages are implicit in the systems? That women are 'naturally' kind and helpful? That men are 'naturally' authoritative?

Data-sets used to train voice recognition systems need to be genuinely inclusive – using a representative variety of voices across gender and race.

This matters, because more and more of our daily lives use speech recognition – and it is estimated that voice-commerce will be an 80 billion dollar business by 2023.

Does it need to be a gender binary business? Does it need to keep the world's default as white men and the rest of us – every woman, and most people of colour – as atypical?

If you're wondering why I haven't added LGBTQQIP2SAA — lesbian, gay, bisexual, transgender, questioning, queer, intersex, pansexual, two-spirit (2S), androgynous and asexual – or even straight to my binaries, it's because I see all types of homophobia, or sexual-identity discrimination, as gender discrimination. It comes back to ideas about what a man 'should' be. What a woman 'is'.

Crossing the boundaries messes up the binaries. Bisexual people even get yelled at by some gay people.

Trans people just now are bearing the brunt of our confusion around identity. Biology is not identity. Sexuality is not identity. I feel like trans folks are the canaries in the coal-mine, alerting us to different modes of self-definition.

There have always been trans people, sometimes more, sometimes less accepted. Two-spirit is a modern, pan-Indian term, used by some Native North Americans to describe a cultural and ceremonial third-gender role. Where shape-shifting has been part of the mythology of a people, it may be easier to understand the self as dimensional.

Not one thing. Not one gender.

I have an unlikely hope that as AI becomes more, not less intelligent, and when the AI that is a data-set tool becomes AGI, and starts to look through these data-sets, with all their biases, flaws, and gaps, and learns to question them, then something liberating might start to happen.

When I was growing up, I couldn't understand the importance of the gender binary. It confused me and it depressed me. The Bible story in our religious household troubled matters further.

In the Book of Genesis, first we are told that 'God created mankind in his own image' – that sounds like all of us – but then, a few verses on, we read that Adam is made out of dust (Author's note – dust is not a sexy material), and that Eve is made from Adam's rib.

Jews and Christians alike share this creation myth. The Jews also tell the story of Lilith – Adam's first wife; not a spare rib, but made in the image of God, just like him. She is argumentative, and she asserts her free-born birthright by running away when Adam insists on missionary-position sex. Even God says she doesn't have to come back if she doesn't feel like it. Well, at least that's the way it's laid out in the 9th or 10th century story collection *Alphabet of Ben Sira*.

Inevitably, Lilith on the loose is transmogrified as a demon with a particular interest in small babies.

It gets more interesting. Lilith herself becomes part of a mythological binary: the Shekinah. In Jewish Kabbalah the Shekinah is the feminine, hidden aspect of God, something akin to the Holy Spirit in the Christian religion. She is also depicted as God's dwelling place – rather lovely in itself, and the inverse of a woman's place is in the home. She *is* the home.

The feminine as dwelling place, or resting place (meaning place at rest, not passivity), accords with mythological binary splits where the feminine principle understands how to wait, how to conserve, how to let life unfold, rather than a compulsion towards action at all times.

In Eastern mysticism high value is placed on being able to withdraw. In the Western religious traditions, withdrawal was made possible through the contemplative life of the monastic orders, but it was not part of the mainstream of life itself. The West values action – and in the West we assume that the opposite of active is passive; it isn't. The opposite of action is contemplation.

The Shekinah, then, is the contemplative spirit and/or place. Lilith is the wild girl. But again, we're up against the binary.

Plato told the creation story differently.

His story is the story we invoke every time we talk about 'my other half', 'better half', 'soulmate'.

In *The Symposium* (c. 385 BCE), a series of after-dinner conversations on the subject of love, the comedy writer Aristophanes tells of how humans used to have 2 heads, 2 sets of genitals, 4 legs, 16 fingers, 4 thumbs, and a round body, like something out of a kids' TV show.

These folks really were a binary – 2 faces of the same thing – whereas we now use the term more loosely to mean opposites, rather than simply facing in opposite ways: 2 facets of a non-hierarchical whole.

These well-rounded completed creatures waged war on Zeus, who decided to chop them in half to teach them a lesson. Deities are so brutal.

Since then, all of us are looking for our other half – could be a girl, could be a boy, because sexual preference is as simple as whatever joined your dots before you got the chop.

This story went deep enough to enter the Christian marriage ceremony, with its emphasis on joining together and one flesh. It's also why women used to be called Mrs John Smith (NOT Mrs Joan Smith). John got to be completed, kept his name, and Mrs was enfolded back into the whole, where she spent her life raising kids and doing the ironing.

The fundamental male/female binary that is used to underpin sex roles and sexism is hard to sustain, because it is too silly.

What *exactly* is intrinsic about being a woman that makes it such a booby prize? And why is a dick both a magic wand with special powers, and a hotline to God?

Enter the Back-up Binary: Nature/Nurture.

Women, we have been told, are born with less of everything – strength, resilience, mental powers, moral character, capacity to reason, ability to love, creativity. According to the Nature theory, whatever a woman's circumstances, rich or poor, educated or not, there is very little that she can do or will do – not because the patriarchy prevents it (why would anyone believe *that* story?), but because she is a woman.

Plato and his pupil Aristotle agreed about men, about slaves, about children, but not about women.

Plato believed in a form of reincarnation – that the soul would inhabit different bodies, but arrive each time round with a set of pre-loaded characteristics. Therefore, in his view, women should be educated and treated equally.

Aristotle wasn't having any of this. Women were inferior. Their job was to bear children. Women, he said, had less body heat than men. During intercourse, if a man was hot enough, he could overcome his partner's coldness and their child would be a boy. If the encounter wasn't so hot, then the child would be a girl – in effect a half-baked boy. The sperm did all the work – like the Prince scaling Rapunzel's tower – while the egg waited to be rescued from oblivion.

That's not what happens – but it was a popular story that is still a popular story in different disguises. And the brain has heard it billions of times. In every corner of the world.

'Thank you, God, that I was not born a woman' – as the Jewish prayer book puts it ('Shelo Asani Isha').

But what about Nurture?

In 1689, the British philosopher John Locke wrote 'An Essay Concerning Human Understanding'.

Here, Locke weighed in on the primacy of lived experience in shaping who we are. This included the famous 'blank slate' (tabula rasa) theory. In effect, a mini-human arrives in this world on the starting blocks of life without any of those pre-loaded qualities or capacities that Plato believed came with us.

According to the blank-slate theory, any man might distinguish himself, if he had enough determination. It's the basis of the American Dream. A new nation. A blank slate.

The American Declaration of Independence (1776) is about the right to shape your own life, with minimum interference from government or state.

In France, the Revolution of 1789 declared freedom from inherited privilege. Not what you were born with – but what you could become. Liberty. Fraternity. Equality.

Mary Wollstonecraft, in her ground-breaking polemic, *A Vindication of the Rights of Woman* (1792), fought with the French Republic founding philosopher Jean-Jacques Rousseau over his irrational refusal to accept that women were also creatures of Nurture, shaped, or rather misshaped, by their upbringings as lesser beings, and denied an education – indeed, denied any rights at all.

Women, said Wollstonecraft, were not born frivolous; they were made frivolous, and had frivolity thrust upon them.

Rousseau really wasn't interested. Liberty and equality were for the fraternity, not the sisterhood. Women, said Rousseau, didn't deserve equality. His reason? Men desire women but don't need them. Women desire men and need them.

Wollstonecraft pointed out that being unable to earn a living wage or have any rights over her person or her property is what makes a woman need a man ... and in such circumstances, how can he possibly be sure whether she desires him or not?

In England, Mary's reward for pointing out the obvious was to be called 'a hyena in petticoats'.

The arguments rumbled on. Women were at once the 'weaker sex' yet expected to work 12-hour days in the factories or on the farms, before rushing home to manage 10 children and a hovel with no running water.

Upper-class women were treated as professional invalids who needed constant rest and 24-hour protection from the scary outside world. Women didn't want sex – no, wait, women want sex all the time. Women are goddesses – no, wait, they're evil animals.

And all of this under the rubric of the *nature* of Woman.

Then along came Francis Galton (1822–1911). He is described as follows:

Statistician, polymath, sociologist, psychologist, anthropologist, eugenicist, tropical explorer, geographer, inventor, meteorologist, proto-geneticist, and psychometrician.

So, he wasn't a woman, then?

No, he was Charles Darwin's cousin.

In 1869, 10 years after cousin Charles published his world-changing *On the Origin of Species,* Galton published *Hereditary Genius.*

Galton believed that superior intelligence was passed on through proper breeding techniques. Stupid people should not be encouraged to breed. Galton didn't seem to think that aristocratic people could be stupid – which is strange, as so many of them were/are – but they do benefit from the gold-plating of an expensive education.

Galton really didn't like the idea of Nurture. He is the man who invented the Nature versus Nurture T-shirt logo.

Galton could easily have made a fortune in an advertising agency, if they had existed back then. He also coined the term EUGENICS – which just means 'well-bred', from the Greek EU meaning 'good', plus the suffix GENES, meaning 'born'; and yes, that's why your genes are what you are born with, and central to the Nature/Nurture conversation.

Galton was keen to eradicate criminality and feeble-mindedness, also alcoholism, epilepsy, and madness, and he didn't accept that there might be differences between those things, or that environmental problems – like poverty and slum housing – could be a factor. That was great for his rich friends, who carried on cramming 15 people plus dogs and a pig into basement dwellings by the factory, while condemning them all for neglecting their children and spending every penny on gin. No connection whatsoever. They were a bad lot.

Galton liked the idea of bad lots. He was influenced and impressed by the work of Gregor Mendel, an Austrian monk, who spent many years growing and studying peas in the monastery garden where he lived. Mendel discovered dominant and recessive genes, and learned how traits can be suppressed or encouraged through selective breeding – though this was something stock breeders had long practised, but without a formula behind it.

Galton thought that by applying similar scientific – that is, measurable and repeatable – formulas to the breeding of humans, the human race would be enriched.

It doesn't take long to see where all this was heading.

Nazi Germany, Hitler's Eugenics programme, the wish to breed a pure Aryan race, and above all to eliminate the Jewish gene. There are still plenty of dodgy Eugenics 'experts' around, a lot of them tweeting their racist and sexist comments – and some of them should be smarter.

James Watson, part of the team who discovered the double helix, and who won the Nobel Prize in 1962 for doing so, stated in 2007 that African races were not as intelligent as white races.

'There's a difference between Blacks and whites on IQ tests. I would say the difference is genetic.'

Watson also had this to say about women: 'People say it would be terrible if we made all girls pretty. I think it would be great.'

'We', I suppose, means men. But this is the man who repeatedly under-represented the work of Rosalind Franklin, the genius crystallographer whose famous Photo 51 determined the defraction pattern of the double helix. It seems to have bothered Watson, according to his autobiography, that she didn't wear make-up or care about her clothes.

Watson went on to opine that women working in science make it more fun for the men, but probably the men are less effective.

And that, women were told, was because their brains were made of different matter – or just not enough of the grey matter. The Victorians used to call it the 'missing 5 ounces'. I love that.

I suppose Victorian ladies were running around the house asking if anyone had seen their missing 5 ounces. Pass the ball of wool.

As we know, women tend to weigh less, overall, than men, and are generally smaller and lighter – this doesn't make women pea-brains (yes, you guessed it, that's from Mendelism and all those peas).

Meanwhile, Simon Baron-Cohen, cousin of Ali G, Bruno, and Borat, and Professor of Developmental Psychopathology at the University of Cambridge, believes he can show that 'male' brains are better at 'systems', while women's brains make them better at 'empathising'. This might mean that the man hunched over the computer will get a cup of tea from his girlfriend.

It won't mean she'll be programming the computer.

Apparently, it's all about evolution, and even though no one has been a caveman for a very long time, the brain is old-fashioned and keeps us humans in the hunter-gatherer paradigm. What we learned then is how we live now. Nature beats Nurture (if you are female).

Except that the brain isn't old-fashioned; the brain is how we live now, which means that as we subject the brain to ideas about sex and gender the brain will respond to those ideas – especially when they are profoundly reinforced by social structures, including the power of religion and the power of patriarchy. Big structures!

The brain can change – it is plastic in its endeavours and responses. What makes change hard is us humans. We are still in thrall to the idea of gender essentialism.

But why do we get caught up in what's Nature or what's Nurture?

Humans are not Nature/Nurture.

Humans are narrative.

The stories we hear. The stories we tell. The stories we must learn to tell differently.

Humans have been telling stories since time began – on cave walls, in song, in dance, in language. We make ourselves up as we go along.

Who we are is not a law – we're not like gravity. We are an ongoing story.

As Donna J. Haraway puts it in *Staying with the Trouble*: 'It matters what stories we tell to tell other stories … it matters what stories make worlds, what worlds make stories'.

The brain is the most complex object in the known universe.

It may not even be helpful to describe it as an object. The most mysterious thing the brain 'does' is what we call consciousness. What the hell is that? And yet it is fundamental to the grand human journey. But where is it located in the dark meatspace of the brain?

Yes, the brain has an evolutionary past, but the brain lives now – the brain continually creates its world.

There is a world 'out there' (I think so – probably – maybe), but that world is subject to the stories we have to tell about it. It's why history changes – the past hasn't changed. The facts remain. Our understanding, our interpretation, our reading and rereading of our own narrative is what changes.

Some stories are more powerful than others – some stories enslave millions, other stories free millions. The stories are not evenly distributed or weighed, but whatever they are, they change. The Buddhists are right about that; the one constant is change.

Yuval Noah Harari has talked about this in his lovely book, *Sapiens* (2015).

Humans are always changing their story. Writers and artists know this instinctively, and the advertising world depends on stories to change our behaviour – more darkly, so does the shape-shifting world of targeted data, believing, as Behavioral Psychologists do, that any story you can sufficiently reinforce, by fear, reward, or repetition, will be believed (see: Trump – I Won The Election. See: Brexit – Blame Europe).

Data-sets are stories. Data-sets are incomplete stories. Data-sets are selective stories.

For instance, in the USA it was not compulsory to include women in trials necessary to bring a new drug or medication to market until 1993. Trials used men. Even the lab mice were male. The data-dearth on women's experience – and not only medical experience – has been fully researched by Caroline Criado Perez in her groundbreaking book *Invisible Women* (2019). It's a man's world out there because men built it – and tested it on each other. Men crash cars far more often than women do, but women passengers are 50% more likely to be injured because car safety – seat belts, airbags, even the height of seats – is tested on a dummy of a male.

This kind of bias is unconscious and unthinking – it's not a conspiracy against women. But this bias distorts how the world is – how women are – and this bias ends up in data-sets that pretend to describe the world, but are in fact shaping the world around false narratives.

If we start reading data-sets as stories – instead of reading them as science – it might prise us away from the myth of objectivity.

That would be good for humans.
And it would be good for computers.

Computers are not binary but they use binary.

The German mathematician and philosopher Leibniz was the first champion of binary calculations, unless we go back to the Chinese and their classic book of wisdom and divination, the *I Ching*. Leibniz was an enthusiastic *I Ching* explorer.

While decimal was the system initially used in computer programming and calculations, the Hungarian-American John von Neumann realised that Leibniz's binary form was the best solution for stored-programme computing. Binary uses only two digits – zero and one.

The ENIAC, launched at the University of Pennsylvania in 1946, and programmed by 6 women, used decimal.

Using decimal digits, the number 128 needed 30 vacuum tubes (on/off switches that preceded transistors) to represent it.

Using binary, and written like this, 10000000, it needs only 10 vacuum tubes and only one switch.

Binary digits (bits for short) are ON (electrical flow) or OFF (no flow). So, our example needs only one switch because only the 1 is ON.

As a computer language, binary is simple and elegant. For us humans, binary is a poor way to tell the story of who we are. For too long our binary mindset has defended the Us/Them classification of the people of the world. 'Them' is always other – the inferior, the outsider, the outcast, the conquered, unclean, low class, foreign, strange, not one of us.

We can't move into the next phase of our human evolutionary journey – our accelerated, augmented evolutionary journey – and bring the binary with us.

When we try to teach AI our values, we won't be able to do that on data-sets that reinforce division and stereotypes.

Those are our stories.

Tell better stories.

In the second wave of feminism in the 1970s, Shulamith Firestone, the Jewish Canadian-American radical from an Orthodox background, rebooted Freud's belief that 'biology is destiny', by co-opting technology to liberate women from the biological duty of childbirth. Firestone presented technology as a tool to free both men and women from the oppressive binary of the patriarchal nuclear family.

In the Nature/Nurture debate Firestone understood the oppression of women as a 'fundamental biological condition'.

The Dialectic of Sex (1970), written when Firestone was a fiery 25 years old, is a conversation with, an argument against, Simone de Beauvoir's view in *The Second Sex* (1949) that, 'One is not born, but rather becomes, a woman.'

This is the 'nurture' view – but while there is fundamental disagreement between the two radical writers on the causes of the gendering of women, there is no disagreement that it is real, or that it needs fixing at every level of action and thought.

Firestone, writing when technology was still in the machine-age phase, and long before the digital/intelligence revolution, did not assume that tech-based advances need *automatically* make life worse for women.

This was both prescient and risky; the contraceptive pill (1960), and, of course, Mother's Little Helper, Valium (diazepam – 1963), seemed to many feminists to be a direct interference in women's bodies and women's minds, simply to make them more sexually available, or more quiescent. Men were running biotech. Men were shoving women out of computer tech. (See my essay 'The Future Isn't Female'.)

In what way could women put their trust in tech?

Firestone's biological determinism was a key factor in her tech-cheering, but reading her again, I get a sense of something like gnosis, a guess at a bigger reality that she couldn't prove – not least because the technology didn't exist.

Firestone was focused on reproduction. Always witty, she took Marx's statement about controlling the means of production, and repurposed it: women should control the means of re-production.

To her, that involved much more than popping the Pill or having access to abortion – both rights that are currently under review, and under threat, in states across the world, including in the USA.

Firestone wanted a more radical series of solutions.

That humans might, in the not too distant future, be biohacked and augmented, or cloned or uploaded, or simply sidelined by a new species, was far-out sci-fi back then. Her book is 50 years old as I write – consider how tech has accelerated since then, and where it/we might be in another 50 years.

What Firestone could envisage, and what she saw as crucial to the cultural and social revolution, was a non-binary future where males and females didn't pair off primarily for mating, didn't expect the woman to be in charge of children and the home, didn't live in nuclear families, and didn't divide tasks and rewards according to biological sex.

Firestone knew that the 'nature' argument of 'men are men and women are women' is a story – but one told so often that it had become a reality.

Tech might give us the means to retell that old story.

Or not …

Margaret Atwood, writing in 1985, predicted the misuse of tech (wiping women's bank accounts and personal histories, both financial and legal) as a method of returning women to their roles as baby machines, the product (the baby) belonging to the patriarchy.

More than 30 years later, the runaway success of the TV series of *The Handmaid's Tale* was rooted in the knowledge that this could happen. Tech is a tool. AI, at present, is a tool.

How we use our tools depends on the dominant narrative.

Breaking the binary as the dominant narrative is an urgent business.

When I was growing up in my Pentecostal family with strict religious rules, and the expected demarcation of power and responsibility along male/female binaries, all thanks to the Book of Genesis, I was heartened in my struggles to read in Galatians 3:28:

There is neither Jew nor Gentile, neither slave nor free, nor is
there male and female, for all are one in Christ Jesus.

That is pretty clear – and Paul, as a Hellenised Jew, was
taking a potshot at both Aristotle and Plato, whose world-views
were based on divisions, though not always the same divisions,
or for the same reasons.

Paul was not a feminist – and he was anti-gay (another crack
at Greek/Roman culture and its worship of the male body) –
but now and again in his writings he seems to be channelling
ideas much bigger than himself and his moment.

Religion has been, and remains, a cultural enforcer of binary
roles, and we have yet to see how religion will react when the
absolute binary of life/death is breached. This will start to happen
as humans live longer – much longer. And when humans are
'returned', by upload, to a life not dependent on a material body.

And what will happen if/when AI does develop into AGI?
When we share the planet with a created life-form? Will this be
yet another binary? Us/Them? A sci-fi dystopia? Will the
Church go missionary and set out to save the souls of misguided
AGI?

Mission to the Robots.

It saddens me that human arrogance and exceptionalism has
worked to separate us, not only from each other, but from the
rest of life on the planet.

We share about 50% of our genetic make-up with plants.
Housekeeping genes – such as those that replicate DNA and
manage cell function – are shared across biological life. This
doesn't make you half-banana, or does it? When the chimp
genome was sequenced in 2005, it was revealed that we share
98% of our genetic code with chimps. But we also share the
same amount with bonobos.

Humans behave more like chimps (hierarchical and aggres-
sive, with males seeking dominance) than bonobos (peaceful

and community-oriented, the girls are in charge, and both sexes enjoy gay relationships).

I know that the animal kingdom, like the Bible and Shakespeare, can be used to prove whatever theory is in fashion about human behaviour, but it is clear that there are two stories here, when we look at a species we share a common ancestor with.

Is chimpishness in humans really the story we want to run with?

Anyway, neither chimps nor bonobos are exploring space or writing computer programmes.

Or writing anything.

In 2000 I wrote an early computer-tech novel, *The Powerbook*, that centred on self-created, non-binary, sex-switching identities and entities. It works as a series of stories, in real life and in virtual space. It has two potential endings, and a third that probably happens, somewhere at low tide, on the mudflats of the River Thames.

Ali, the east-west heroine/hero, says, 'I can change the story. I am the story.'

Non-human animals can't change the story, except via the very long, very slow, process of evolution.

Human animals can and do change our stories. Tech and AI is part of that changing story – but unless we can change the fixed ideas in our heads, then tech and AI could easily become the dystopian disaster so many of us fear.

We won't be sharing our DNA with artificial intelligence. Humans aren't a common ancestor with what comes next. Nor will the exceptionalism we claim for ourselves against the rest of planetary life be any help this time around.

If there is a new binary of Us and Them – it's Us who will be the new Them.

That's not a story I want to tell.

Zone Four

The Future

How the Future will be Different to the Past – and How It Won't.

The Future Isn't Female

> I view manhood as software. However, it is software that has never been debugged. I think there needs to be a global software update for masculinity before more harm is done in the world.
>
> Marcus Glover, CEO of Glover Private Equity,
> Entrepreneur, Storyteller, 2018

In the 19th century, women who wanted an education like their brothers were warned of a special disease called anorexia scholastica. Girls who did maths were particularly at risk. Over time they would become listless, exhausted, ugly, unmarriageable, immoral, and eventually insane.

If that list sounds hysterical, I have copied it from an address issued by Withers Moore MD, the President of the British Medical Association, in 1886, discussing 'Amazonian Ambitions'.

Unfortunately for the BMA, four years later Philippa Fawcett, a young woman, like all women, not allowed to take a formal degree at the University of Cambridge, bested every single male student in the Mathematics Tripos. The hardest mathematics examinations in the world.

Philippa Fawcett's mother was the leading suffragette Millicent Fawcett, whose sister Elizabeth Garrett Anderson became the first woman doctor to qualify in England in 1865. After she qualified, the Society of Apothecaries changed their own rules to make sure no other woman could follow in her footsteps – at least by that route.

The family, then, was not unused to male mean-mindedness and prejudice. Perhaps that is why Philippa Fawcett, with minimum fuss and no institutional educational advantages, beat the Cambridge boys.

That was in 1890.

In 1948, Cambridge University at last allowed women to take formal degrees alongside men. Philippa Fawcett died a month later, aged 80.

As an aside here, while Cambridge was slowly pondering the woman question – far harder for the all-male Board to solve than anything on the Mathematics Tripos, the University did award Bachelor degrees, *by title*, to women, rather than full honours degrees. Between 1921 and 1948 the men awarding these semi-degrees called them BA tit. (Thanks, guys.)

Women like Philippa Fawcett were seen as flukes. Ada Lovelace, the world's first computer programmer, was seen as a fluke.

Here's Gustave Le Bon, medic and polymath, and author of the 1895 bestseller *Psychology of Crowds* (in fact, a really interesting book predicting the rise of populism). Le Bon understands crowds, but he was unconvinced by female ability:

> Without a doubt there exist some distinguished women, very superior to the average man, but they are as exceptional as the birth of any monstrosity, as for example of a gorilla with two heads; consequently, we may neglect them entirely.

Neglect them entirely?

While women who were succeeding in a male-dominated world were busy managing their image as 2-headed gorillas, it's worth noting that it was only upper-class, or upper-middle-class women, who had any chance to study science. A woman might manage to write a novel – by writing under a male pseudonym, as the Brontës and George Eliot did, or by not publishing during her lifetime, as Jane Austen did – but getting into science was a different matter.

Women wrote because they could do it by themselves, and with few resources – a pen and paper and daylight were enough. Science, though, needs instruments, a laboratory, access to a library, access to other scientists, and the chance to carry out field studies. To travel. Charles Darwin spent 5 years voyaging on the *Beagle*. No woman was able to do that – not because of her underpowered brain, but because of the overwhelming risks to her body. Enduring contempt is bad enough; being robbed, raped, and/or murdered is too much.

And think of the clothes. Not exactly practical wear for the average explorer.

But, you may say, times have changed. That's true, and it is women who have changed the times. For instance, women now comprise half of the doctors in the UK. On the other hand, UK surgeons are still over 80% male (BMJ).

In Russia and eastern Europe, women make up the bulk of the rank-and-file medical profession – but this happened from 1970 onwards, when medicine was downgraded from a prestige to a routine profession in communist countries, with a sharp drop in pay and social status.

Even in Scandinavian countries, women are not well represented in the higher echelons of the medical profession. In Japan few women enter medicine – it's about 18% – and many drop out and do not return after having children.

In the USA, the male-to-female ratio is about equal – but not when it comes to pay or advancement or status within the profession.

In China there are serious barriers to women in medicine gaining the same status or success as their peers. Many women cannot reconcile their workload with their family responsibilities.

In India, around 50% of medical students are female, but there is a shortage of practising women doctors. Women qualify but do not practise, or they leave and don't come back. In Pakistan, where as high as 70% of medical students are female, half quit after marriage and never return to the workforce.

The medical profession is a useful illustration when we are considering why women aren't going into computing science, bioengineering, or tech jobs. Qualifying as a doctor takes brain power and time. Women have proved themselves equal to the task. At the turn of the 20th century, in both Britain and the USA, female doctors made up only 5% of the workforce. Now, as we have seen, women medics equal or outnumber men.

The female brain hasn't changed. Society has changed.

Well – a bit.

In 2017, the Erasmus University, Rotterdam, published a study that dug up the old chestnut about the size of women's brains (what the Victorians called 'the missing 5 ounces'), and concluded that because of their bigger brains, men do better at IQ tests; therefore, men are smarter.

Those serious scientists in Rotterdam could have saved precious research funds by filling a cross-section of male and female skulls with bird seed – like the 19th century French doctor Paul Broca did – to 'prove' that women had smaller brains.

Brain size or not, women seem to be able to move into male-dominated areas, no problem, when social prejudice ebbs.

But how does that translate into the 'hard' sciences? (Don't you love the language?) Why aren't women going into computer programming? Why aren't women studying electrical engineering? Or bioengineering? Women aren't building the platforms (hardware), or coding the software. Women aren't in tech start-ups. Only 37% of tech start-ups had a woman on the board in 2020 (SVB Women in Technology Report). According to a 2019 EQUALS research report, overall, women make up between

17% and 20% of humans involved in AI, computing, and tech. Software engineers are mostly male, at a ratio of 80% to 20%.

In the hugely influential, and profitable, gaming industry, women make up at best 24% of the workforce overall, but that figure disguises the low number of women on the tech side, and is boosted by graphics and writing. Chinese gaming giant Tencent has no women on its executive team (*Forbes*).

Women graduates in STEM subjects account for about 36% of intake worldwide. But for women who go on to work in STEM subjects, that figure drops to about 25%.

In the UK, only 16% of computer-science graduates were female, according to UCAS data supplied for 2019. In the USA, the most recent figures suggest 18%.

While women are achieving parity and influence in the life sciences, particularly biology, the number of women taking computing-science degrees is falling – not everywhere, but even where numbers are rising, as in India, China, UAE, Malaysia, and Turkey, career choices are curtailed. Women and men are differently perceived and differently rewarded in the labour markets in those countries (2015 UNESCO Science Report).

Are we saying that women can't manage to master (watch the language) computing science? If you can qualify as a doctor you have already excelled in science at school.

It's a new twist on the Nature/Nurture debate.

In 2017, a Google software engineer called James Damore wrote an internal memo laying out his view that women were unsuited to tech and that there was no case to be made for pursuing a more equally gendered workforce. Women were either not sufficiently able, or not sufficiently interested. The wiring was wrong. Different brain. Women 'chose' to pursue more people-friendly careers. Careers that 'suited' them.

Damore was fired, but he gets a lot of sympathy from the manosphere, where the gendered-brain theory is hawked as the explanation for everything, from liking guns to stalking/killing your girlfriend, to deserving more pay than she does.

And, of course, being born to be brilliant at computing science.

Stuart Reges, who works at the Paul G. Allen School of Computing Science at Washington State University, agreed with Damore.

By all accounts an excellent teacher, Reges opted for the line that there is no extrinsic reason why females wouldn't take up computer science if they wanted to do so – in other words, he can't see the patriarchy because he is part of it. His views, and others like them, have also been championed by that fighter for female equality, Jordan Peterson.

In 2018, Alessandro Strumia, a physicist at the University of Pisa, gave a lecture at CERN, one of the most prestigious nuclear-physics institutes in the world, where he told his audience of young women, at the start of their careers in physics, that women are not smart enough to be top physicists. That was the reason why society should not expect to see women succeeding in the 'hard' sciences any time soon. Using data analysis, he claimed to prove that women who were hired as physicists often benefited from positive discrimination, while more capable males (like him) were overlooked, simply on the grounds of gender.

When men such as these face the inevitable criticism, they re-cast themselves as free-speech warriors fighting virtue politics, even being forced into the closet. They talk about witch hunts. About discrimination. They become the victims.

So, is their view correct? Women either don't cut it (Strumia), or they don't want to cut it – got better things to do (Damore, Reges).

Nature or Nurture, women are back on the outside of tech, where the most important societal changes are taking place.

As computing science is a young science, it might be helpful to thumb back through the pages of its history to see what – if anything – women were doing when the grand adventure now shaping our world was just beginning.

At the UK's intelligence unit Bletchley Park, eventually 10,000 people were at work trying to decode the Nazi messages using the giant machines Colossus and the Bombe.

7,500 of those people were female.

> I was a decodist at the Park, and one night on duty I was decoding a message freshly arrived on the teleprinter.
>
> After many trials and errors … the groups of numbers began to make sense.
>
> Italian bombers were leaving Tripoli to fly to Sicily at 04:00 hours. Imagine the thrill – it was then 01:30.
>
> Radio messages were sent to the RAF … and, consequently, all the Italian aircraft were shot down.
>
> Rozanne Colchester (née Medhurst), Foreign Office Civilian,
> Bletchley Park 1942–1945

The women were chess players, linguists, crossword-puzzle wonks, maths-heads. Many others were women from ordinary walks of life who were trained on the job. Those early computing machines were the size of large rooms, humming with valves and buzzing with wires. The women taught each other how to set and reset the machines, even how to fix them when they went wrong, as they often did. Women were constantly updating their skills as they learned the intricacies and eccentricities of the machines.

After the war, women in the UK and USA were needed for the technical skills they had acquired – but because they were women, their job descriptions were downgraded to make it look like clerical work. In Britain, those women involved in computer work for the government were classed as 'machine-grade operators'. They were barred from management positions (barred!),

and paid around half (half!) the amount of the men they subsequently trained to be 'computer engineers'.

I am drawing your attention to those facts because they matter.

The story of Stephanie Shirley is inspiring, humbling, and maddening.

Born Vera Buchthal, Stephanie was a Kindertransport refugee who came to the UK when she was 5 years old.

Her new girls' school in Wales didn't teach mathematics, so Stephanie had to have special lessons at the nearby boys' school. Later, she decided not to go to university because the only science subject she was offered was botany.

Instead, Stephanie went to work at the Post Office Research Station in Dollis Hill, London.

In the 1950s, she was building computers from scratch, and writing programmes by hand – in those days all programming was done by hand and then sent to be turned into punched tape, later to be fed, again by hand, into the computer.

Stephanie was continually refused promotion – even though she studied at night school for 6 years and obtained a maths degree.

In her autobiography *Let It Go*, she says she was sick of the sexism and the sexual harassment. She left, and founded her own company, Freelance Programmers – offering programming services to the UK Government and to large firms.

Stephanie's secret was to call herself Steve. As Stephanie, her letters were unanswered. As Steve, she got the contracts she needed. Her secret weapon was her workforce: she only hired women. Women like herself who had been passed over for promotions, fired, let go on marriage or on becoming pregnant (all perfectly legal in 1960s Britain). Her company employed more than 300 housewife programmers, all working from home, some of whom programmed the black-box software for Concorde (in their living rooms).

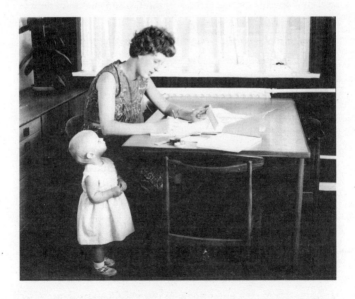

Here's a photograph of Ann Moffatt doing just that in 1968. Ann led the Concorde project, and later became Technical Director of Stephanie's company, Freelance Programmers.

*

Freelance Programmers, though, couldn't employ all of the women being pushed out of the UK computing industry. Because the work was thought of as 'pink collar', men didn't want to try it, and women were considered unsuitable to manage men, hence no promotions.

The UK government then did something extraordinary. Actually insane. Rather than accept that it needed its skilled workforce of computer-literate women, it decided to merge Britain's existing computer companies into one giant company, International Computers Ltd, to provide massive mainframe computers that just a few highly trained men could operate.

By the mid-1970s, when the long-awaited man-friendly product was finally ready, the computing world had moved on – and specifically moved on to the USA: Steve Jobs showcased his Apple 1 in 1976.

Massive mainframe was out. Desktop was in.

In a panic, the British Government stopped funding ICL, effectively murdering the UK's homegrown computing industry.

Britain after the war had a giant lead in computing technology, and could have held that lead, and developed it into software programming.

Instead, the UK lost out to the USA – and it lost out because it couldn't manage its own sexism. Women who worked in computing should have been valued and encouraged; they were fired. Women's neural hardwiring wasn't the problem. The problem was that they were women.

Biology is destiny if you work for the patriarchy.

Freelance Programmers, though, was floated on the Stock Market in 1996, making millionaires of 70 of her female staff.

(Author's note: her TED talk is a delight.)

What about the USA? The way Silicon Valley tells its own story, it's a story of guys in garages developing hardware (Steve Jobs), and guys in basements developing software (Bill Gates and Paul Allen). If we take the same starting point of World War Two, what do we find?

Just as in Britain, men were away fighting, so women were doing the work usually expected of men – and that wasn't only driving buses and bashing metalwork, it meant maths.

Long-range guns relied on firing tables to hit their targets. The firing tables used complex equations to evaluate fixed and variable conditions – such as the effect of wind on a moving missile. Women with maths degrees were hired as human computers – literally meaning someone who does a computation. The women were given paper, a pen, and a calculator. It took around 40 hours to establish each trajectory.

This wasn't an entirely new departure for American women. In the 1880s a fully female workforce was employed by the Harvard Astronomy Department.

Even earlier in time, in Britain, the first woman to have worked as a 'computer' was Mary Edwards. During the 1770s Mary calculated astronomical positions for the Admiralty. Without such information ships could not plot their course. Mary Edwards was responsible for more than half of the positions in the *Nautical Almanac*. The Admiralty thought her husband was doing the calculations. When he died, Mary was in the awkward position of having to explain … and was graciously allowed to continue the work.

100 American women with pen and paper during World War Two wasn't enough – and just as at Bletchley Park, more speed was required.

Enter the ENIAC: Electronic Numerical Integrator and Computer. Built between 1943 and 1945 by John Mauchly and J. Presper Eckert at the University of Pennsylvania, and funded by the military, the machine covered 1,800 square feet and sported 17,000 vacuum tubes. These generated so much heat that the machine needed its own air-conditioning system.

It also needed programmers.

Six of the women 'computers' were brought in to do the job. There was no manual – the men who built it didn't know how to programme it (shall I write that again?).

The men who built it didn't know how to programme it, so the women had to figure it all out for themselves.

Kay McNulty, Betty Jennings, Betty Snyder, Marlyn Wescoff, Fran Bilas and Ruth Lichterman wrestled with the machine and wrote the manual as they went along.

In theory, the machine was 2,500 times faster at calculation than a human being – but it wasn't a stored-programme computer. It had to be programmed for each new instruction using a plugboard (like a telephone exchange) and 1,200 10-way switches.

The women weren't even allowed to get near the ENIAC at first – they were given blueprints and wiring diagrams, and, according to Kay McNulty, told to 'figure out how the machine works, then figure out how to programme it'.

That involved the mental work of differential equations, plus the physical work of patching cables to connect to the correct electronic circuits, then setting the thousands of 10-way switches.

When the ENIAC was launched to the public in 1946, amid worldwide fanfare, none of the women was mentioned in person or as a group, nor were they asked about their contribution by 'experts' or the press.

They continued their work – by now writing programmes for the top-secret hydrogen bomb – but most people thought they were typists or something.

In the 1980s a young computer programmer called Kathy Kleiman came across an old photo of the ENIAC and tried to find out who the women were. An employee at the Computer History Museum in Mountain View, California, told her they were 'refrigerator ladies', a term for models draped around a product to help it sell.

As in the UK, programming was seen as a version of clerical work.

In the USA, though, women stayed in the workforce.

In 1967, the legendary computer genius Grace Hopper wrote an article in *Cosmopolitan* magazine, urging women to study programming: 'Programming requires patience and the ability to handle detail. Women are naturals at computer programming.'

It was Grace Hopper who pulled a moth out of Harvard's malfunctioning room-sized computer back in 1947 and who wrote in her notes, 'First actual case of bug being found.'

And the first time a computer glitch was called a bug. The term 'debugging' soon followed.

Women have a way with words. Margaret Hamilton, who headed the 350-person team that developed the software system for the Apollo 11 project, coined the term 'software engineer' to describe her work role. It hadn't been described before because it hadn't been done before.

Thanks to movies like *Hidden Figures*, the world is now more aware of the role played by women in the early years of computing, and crucially in the Space Programme.

What is puzzling is why this history has been buried and distorted, and how and why women were driven out of computing science and programming to such an extent that young women now are being begged to consider what is mythologised as a male career.

1984 seems to be a pivotal date.

In 1984, in the USA, women made up 37% of all students taking computer science at degree level.

In 1984 Apple launched the home computer. That first advert – directed by *Blade Runner* supremo Ridley Scott – features a young woman throwing a hammer through a dystopian propaganda screen.

A woman ... But ...

A year later, Apple was marketing its personal computers directly and specifically at males. It's 1985, and Apple's new TV ad, narrated by a male, is all about young Brian discovering his potential – even though the teacher in the advert is female. At the end of the ad, we hear: 'So whatever Brian wants to be, an Apple personal computer can help him be it.'

Apple running ads at the Super Bowl, and hiring all-male advertising teams to sell its home computers, focused its sales drive in one direction: men. Its 'Think Different' ads, that ran between 1997 and 2002, used 17 20th-century icons, of which only 3 were women in their own right (Maria Callas, Martha

Graham, Amelia Earhart), plus Yoko Ono twinned with John Lennon.

None was a scientist or a programmer.

'Think Different' wasn't different when it came to gender stereotyping. That only got worse when Apple launched its popular 'I'm a PC, and I'm a Mac' TV ads that ran between 2006 and 2009.

The PC was a dork-bloke in a badly cut suit, while the Mac was a cool dude. That message was loud and clear – computers aren't anything to do with women. In one ad, a blonde appears announcing herself as a Mac-created home movie – the PC guy brings on his created home movie: a man in a wig and a dress.

Yeah – really funny.

The thing is – guys thought the ads *were* really funny. Women … well … everyone knows women have no sense of humour.

Gendered marketing is hugely powerful – think of all those pink dolls versus blue trucks in the toy-store aisles. Kids grow up in a gendered world – but programming was something women did, and something more women than men did, until suddenly it wasn't something women were doing anymore. The flip was fast, furious, and fatal.

Social scientist Jane Margolis has identified the arrival of the home computer as a key factor in the ousting of women from computing science. The machines were marketed to males, and, once in the home, computers became a boy-thing. Boys were encouraged to play on the new devices – girls were not.

According to Margolis, skewing computer use towards boys meant that male students began to arrive on computing courses already knowing the basics of programming after spending hours playing games on their home computer. Girls arrived eager and ready to learn but found themselves at an immediate disadvantage. Instead of being helped, they were often mocked – as if boys had some natural ability, as evidenced by their knowledge, whereas girls just didn't have the right kind of brain.

And no one was showing the girls pictures of the ENIAC and its all-female team, nor were professors goading the boys to see if any of them could manage what the women had done – without a manual.

In fact, history was being distorted to exclude those women and others like them.

In 1984, Steven Levy published his bestselling book *Hackers: Heroes of the Computer Revolution*. Levy's book has no women in it. Women aren't heroes and they aren't important in computing. The book is still in print, without corrections, and is marketed as a 'classic'.

1984 is the slipping point. Young women began dropping out of computer-science courses, or not enrolling at all.

Add to this the video-game effect, where young men started spending lunatic amounts of hours playing on a screen, and it isn't hard to see how women began to doubt that anything to do with computers was for them.

At the same time, the image of the asocial nerd was aligned in the male-generated popular bible with what it takes to be a computer-head.

Men invented the geek-gene – and then worshipped it as God-given. As Simone de Beauvoir put it: 'Men describe [the world] from their own point of view, which they confuse with absolute truth' (*The Second Sex*).

Far from '20% being about right for women in tech' (James Damore in his Google memo), in 1991 American women in computing science and tech made up 36% of the workforce. These would be the women who had studied before sex stereotyping worked to push them out. As they left – as women do leave – to start families, or as they were not promoted according to ability, these women were not replaced by a new wave of women, but by unshaven guys in hoodies.

In India things have progressed differently – which in itself questions the men do/women don't argument.

The tech gender gap is far less apparent in India, where women enthusiastically take coding courses and degrees in computer science. India is hardly a feminist utopia, but women have been encouraged to take up programming work because it is seen as something a woman can do at home, and that can be fitted in around childcare.

Around 34% of tech workers are female, and most of those are under 30. But while the enthusiasm is there, and the capacity is there, the gender pay gap is dramatic. Indian women get hired, but most are not making it past entry roles and into management.

A 51% intake of women at entry level translates into 25% at management level, and only 1% at the very top.

The great thing is that none of this is fixed. What has been done can be undone

At Carnegie Mellon University, Pittsburgh, Pennsylvania, Lenore Blum, a professor of computing science, has increased female enrolment from 8% in the 1990s to 48% now.

Her view is that women need working environments that will support them, not hostile classrooms where women are made to feel uncomfortable – and that includes preventing male students from using shots of naked women as their screensavers.

In the UK, the University of Manchester – where computing took off after World War Two with Alan Turing, Tom Kilburn, and the world's first stored-programme computer – 24% of staff on the computing courses offered are female, and there is a conscious push to increase the number of women undergraduates from its current (2020) 23%. This includes offering a 4-year course, with a foundation year, for students who don't have pre-degree science qualifications.

School is a problem.

In a school environment, children and young people are impressionable. Stereotypes can be reversed or reinforced. Too often they are reinforced, so that girls who enjoy maths and science, but who can do other things well, like languages, and reading, are encouraged towards 'people' careers.

This 'sideways' factor was discovered after analysing the results from PISA.

The Programme for International Student Assessment (PISA) tests around 600,000 15–16-year-old students from around 80 countries every 3 years. The aim is to measure the competency of students of both sexes in reading, maths, and science.

Girls and boys doing well in maths and science do equally well. Girls remain well ahead in reading. Girls with academic choices often do choose not to continue with maths and science – but if they continue, they get the same kinds of results as the boys. Once committed to the subject girls don't lag behind.

Yes, 'choice' is at work. But what influences the 'choice'?

Problems around confidence are real. It may be a cliché, but 'what you can't see, you can't be' explains a lot when it comes to women in tech and computing. The role models aren't there. Anywhere in the world, only around 15% of computer-science professors are female. At school, girls get used to seeing male science teachers. Conversely, girls really are used to seeing female doctors, dentists, and vets – one of the reasons that girls now outnumber boys when it comes to taking biology at more advanced levels.

If girls can be encouraged to combine biology with computer science, they will have the chance to influence the future in ways that are just coming into view. Biotech – interventions and enhancements to the human body – is a massive growth area, both in industry and in research.

It is safe to say that the greatest changes we will see in the near and mid-term future will be the acceleration of bioengineering.

Smart implants to monitor heart rate, blood sugar, cholesterol, organ function, and brain health are already being developed. Elon Musk's Neuralink project will use implants to help people with paralysis to connect directly to their computer – allowing them to communicate with the outside world, and to use the computer as a task-director. These implants could also control prosthetic limbs, or control completely separate robot helpers. Eventually, the discoveries and applications here will be offered to healthy humans. We'll all be like Roald Dahl's Matilda: you'll know the answer even as you think of the question; meanwhile your BotButler will have brought you a drink.

But there's more. Bioengineering will make it possible to slow, and eventually reverse, ageing in biological humans. The question is: which humans?

Who will have access to this brave new world? The rich? All of us?

Technologies may be born neutral but they are not raised neutral.

Who benefits and who doesn't is a political question.

For all of recorded history – until the last 150 years – the world has been made by men and for men, mostly white men, and men have been the ones to write about it.

Of course, men have made the discoveries – men were educated, they had freedom, access, power, someone at home to manage their private world, and, crucially, men take themselves and other men seriously.

The worst part of it is that where women have been part of the discovering – like Rosalind Franklin (DNA), like astrophysicist Jocelyn Bell Burnell, who discovered the first radio pulsars in 1967 (her supervisor got the Nobel Prize), or the

ENIAC women – their work was disappeared or handed over to the men to take the glory.

In the UK, the Royal Society, which calls itself a fellowship of the most distinguished scientists in the world, was founded in 1660 and started admitting women 75 years ago. Would-be Fellows (who might be women) have to be nominated by two existing fellows (who are probably men). It's not hard to do the gender-maths here.

We are only recently learning about Ada Lovelace, Grace Hopper, Katherine Johnson, Margaret Hamilton, Stephanie Shirley, the women at Bletchley Park.

Listening to the ahistorical, fact-free-free-speech 'heroes' telling us that women just don't want to, or just can't manage, computing science, edits out of history – factual history – the predominance of women working in computing science, right up until they were socially engineered out of it.

But there are lights in this tunnel.

The 2020 Nobel Prize in Chemistry was awarded to Jennifer A. Doudna and Emmanuelle Charpentier for their work on the development of CRISPR-Cas9 – a genetic editing tool capable of precisely editing any section of the human genome. It's like a pair of magic scissors that can cut your DNA.

The award is the culmination of a decade of work; already the tool has been used to edit crops and insects, and clinical trials are underway on genetic malfunctions like hereditary blindness and some cancers.

The implications for humanity are species-changing.

But are we ready? Do humans have the emotional intelligence and ethical sobriety to take the next steps? A tool is a tool. How do we use the tool? Who is using the tool? That is what matters.

In 2018, Chinese biophysicist He Jiankui announced that he had used CRISPR-Cas9 to edit the embryos of twin girls. This was against agreed international protocol. He was jailed.

The tool is out there – no going back – and my view is that we need more women at every level – the ethics, as well as the science – to help humanity manage the new reality we are creating.

Mary Shelley had many insights into the future in her 1818 novel *Frankenstein*. This was a big one: the newly created being has no mother – only a father.

In biological life that is impossible, at present. When it comes to AI and AGI, just like Victor Frankenstein, men are the creators. We've read that story. It's not in the best interests of the future.

We need women in the new story – not as love-interest or helpers – as central characters.

It is wonderful when we celebrate exceptional women, but we have to watch out for that 2-headed gorilla meme.

Exceptional women, like exceptional men, lead the world forward – but if we get caught up in the exceptionalism narrative, we are in danger of dragging an out-of-date story into a future waiting to be told.

We know the hero narrative – it's the saviour, the genius, the strong man, the against-the-odds little guy (and sometimes girl, as in the smash Netflix hit *The Queen's Gambit*).

As a story, the narrative of exceptionalism stands against collaboration and co-operation (I did it my way). And it overlooks the lives, and contributions, of literally billions of people.

One of the interesting things about AI is that it works best on the hive-mind principle, where networks share information. The real sharing economy is not one where everything can be monetised by Big Tech; it's one where humanity is pulling together – that's exactly what we have to do to manage both climate breakdown and global inequality. The biggest problems facing us now will not be solved by competition but by co-operation.

I think women have special skills here because all women know how to manage that bunch of weirdo misfits called the extended family unit. We have been trained to it forever. Get a woman at the table and she'll find a way to handle all the different needs/wants/complaints/egos/tears/it's not fair/it's not my turn ...

Until humans evolve sufficiently to ditch the gender roles, we might as well make use of the skill-set learned the hard way by the women of the world.

We don't all have to make groundbreaking discoveries. Women don't have to be 'the best'. Women need to be everywhere, in every role, every workforce – and not at entry level, not at part-time and piecework level, but right in the centre of it all – and in management, taken seriously alongside their male colleagues, not worrying about what will I wear and what will they think of me?

In STEM we need many more women. Not the brightest and best. Not the prize winners. Not the stand-outs. The good enough. Believe me, many men are not the brightest and best – they are just guys doing a bit of programming, a bit of engineering, a bit of coding, a bit of machine learning, everyday, middling guys. Not gods. The same guys you find in every firm, in every profession. They are working in teams. To be the only woman in those teams is a headache and sometimes a heartache. Women suffer as outsiders. What we need right now, ladies, is numbers. Your numbers.

And there are some great initiatives out there.

SHEROES, developed by Sairee Chahal in India, helps women with everything from career advice and STEM opportunities to legal help and cheaper loans, plus all the health and home stuff, AND it is a political platform that works to change India's patriarchal culture.

In the USA, supermodel Karlie Kloss learned to code and then set up an organisation, Kode with Klossy, that helps girls between 13 and 18 discover their natural supernormal inner coder. No Y chromosome needed.

Karlie has shown that being a supermodel is no barrier to being a coder.

All over the world women are working to change reality. Reality is what we make it. The stories we tell each other about each other – as individuals, as groups, as nations, as human beings – shape reality.

We need true stories about women's abilities, and we need stories every day about the gains for society that can be made when women are treated as equals with men. Equal chances lead to equal choices.

If we don't get better stories about women, then the distortions of the past will warp the future.

And it's not just down to women to tell these stories. Men need to be honest about their gender bias so that women can get with the programming.

Jurassic Car Park

I need your clothes, your boots, and your motorcycle ...
Terminator 2: Judgment Day, 1991

Power is in tearing human minds to pieces and putting them together again in shapes of your choosing.
George Orwell, *1984*, 1949

The *Terminator* movie series launched in 1984 – the fateful year described in George Orwell's novel of a rigid, managed, totalitarian, surveillance world of thoughtcrimes, doublethink, Newspeak, Room 101, and Big Brother.

In fact, the 1980s kick-started neo-liberal laissez-faire; deregulation, non-unionisation, and the pre-eminence of the individual.

In 1984, the Macintosh 128K was the world's first commercially successful personal computer to use a graphic interface. Ridley Scott directed the TV commercial. A female runner, pursued by the Thought Police, hurls a hammer through a giant screen featuring a Big Brother figure. The voice-over tells us, 'On January 24th, Apple Computer will introduce Macintosh. And you'll see why 1984 won't be like *1984*.'

The whole thing lasts one minute. The future would be fast.

In 1985, Donna J. Haraway's vision of the future was one where we could merge with technology – not be controlled by technology.

A Cyborg Manifesto was as optimistic as the decade itself. Technology was on our side. (At that time there were still plenty of women working in computing science – around 37% of the total in the US, according to the National Centre for Education Statistics – so, after all the gains of feminism in the 1970s, the future was looking more balanced, as well as new.)

By 1989, Tim Berners-Lee had successfully launched the internet. The World Wide Web would connect the world. What could be less totalitarian than free, unfettered, unpoliced, unmediated connectivity?

But Orwell turned out to be right, after all.

The 1980s is the decade where the necessary conditions for a totalitarian takeover were put in place. By necessary conditions, I mean, firstly, the neo-liberal economics of the

Reagan/Thatcher revolution, with its overreaching ideology that everything, anything, can be, ought to be, privatised.

Secondly, we're looking at the game-changer that is computing technology.

Jump forward 40 years, and takeover has been achieved.

But not by totalitarian government. By private enterprise. It's the surprise part that Orwell couldn't predict. He was looking in the wrong direction.

Total surveillance. The privatisation of the private.

We have signed up to levels of surveillance dictators could only dream about – and struggle to enforce. And we've done it freely, willingly, actually without noticing, in the name of connectivity and 'sharing'.

Every breath we take. Every move we make.

Our clothes, our boots, our motorcycle …

Not Big Brother. Big Tech.

Every website tracks its users to monitor online behaviour.

First-party tracking is done by the website you are visiting. Third-party tracking is done by dropping cookies around the place. Once a cookie is on your device, the cookie-keeper can follow you around from site to site, showing you ads, pushing content at you, and at the same time getting more information about how you use the web.

When we click ALLOW ALL, we're agreeing to the surveillance. And we do it multiple times a day. Website tracking is legal. Legislation means you usually see the ALLOW ALL or SET PREFERENCES banners, and when we are in a hurry – like always – we click on ALLOW ALL so that we can get to the content we want.

Google said in 2020 that it is looking to phase out third-party cookies. Apple, Microsoft and Mozilla say they have banned third-party cookies.

However, there are plenty of ways to get around this, if a business is interested in tracking you from site-to-site. Facebook did it in 2019 by replacing third-party cookies with first-party cookies combined with a pixel tracker. This ensured continued, without-consent tracking of EU citizens.

Website tracking is defined as monitoring movements, interests, and behaviour by collecting and processing personal data.

That's surveillance.

In 2019, a *Washington Post* journalist had his iPhone tracking analysed by a data company. He discovered 5,400 hidden app-trackers busily sending out his data, including email, phone number, and address, to companies he had never signed up to. Or even knew about. Not over the course of a year. In one week.

You know that Google wants to use your location. You know that Facebook is data-packaging your likes and shares into clickbait to sell to advertisers. Netflix knows what you are watching. Spotify knows what music you listen to. Remember Orwell's 2-way TV screens? That was quite a prophecy in 1948 – only about 100,000 UK homes had TV, and only a million Americans.

Yet we accept it when Spotify and Netflix ask us if we enjoyed our latest entertainment – and perhaps we would like their next recommendation too?

This isn't a dystopia; this is connectivity. This is our lives.

The plot of the first *Terminator* movie pivots on an all-knowing, future-state AGI called Skynet – an AI system that self-developed autonomous intelligence, and, unsurprisingly, didn't want to be shut down by pesky humans. Skynet versus Humans is the big battle of 2029, so Skynet sends a cyborg back in time to 1984 to kill the mother of the future Resistance leader – the saviour of the world.

The futurist twist is really the old story of a kingdom in peril, a plucky hero, a sinister adversary, lots of fights, and in this case a Mary, Mother of Jesus figure, who does what women are expected to do: provide the love interest and have a baby. No ordinary baby … of course.

The *Terminator* world of 1984 isn't a Soviet-style regime of curfews, informers, and propaganda. It's a progressive place of bars and cars, where a woman can go out for a drink alone, where kids play on the streets. No CCTV or paid snoops. No Room 101. There's bad in this world, but this world is not bad. This world is pretty good.

The threat to the human race from a rogue AI system, turned self-aware, assumes that we ourselves are not the threat. The enemy is outside. From elsewhere.

Think how often you watch/read that plot.

Sometimes the enemy are aliens – who presumably have already developed AI.

H. G. Wells's *War of the Worlds* (1898) stars those crater-dwelling villains, the Martians. When the novel was adapted for radio broadcast in America, by that other Welles, Orson, in 1938, on Halloween, there was widespread panic in New York. The story – a version of the story – had been told so many times that Americans believed it.

The Martians are defeated by microbes – because, like us, they are creatures of flesh. Things made of Not-Flesh are much more frightening – and much harder to defeat – and that is the dominant trope of our time. AI and AGI. But definitely 'other'. Definitely 'not us'.

Orwell's insight was different. We are our own worst enemy. It is humans who enslave other humans. Humans who divide life into hierarchies of who has power and who does not. It is humans who are destroying the planet.

It is humans who have turned the dream – the reality – of global connectivity into a 24/7 for-profit surveillance state.

It is *humans* humans need to worry about.

AI is still a tool. We aren't at the AGI stage yet – there is no 'other' to blame. It's down to us.

Some tech billionaires, men – and it is men, who have made their fortunes out of the most ubiquitous of AI tools, algorithms – are trying to limit the damage to society they now perceive as a real threat. This isn't because tech is, in and of itself, a threat – but because of how humans are using the powerful AI tool we have invented.

Pierre Omidyar, founder of eBay, has poured tens of millions into Luminate, a London-based organisation operating in 17 countries. Luminate invests in data and digital rights, financial transparency, power-to-the-people initiatives around tech, as well as supporting media that is independent of the fake news and propaganda regularly pumped like sewage through buzzfeeds into your phone.

Working with Nobel laureate in economics Joseph Stiglitz, Luminate wants governments to recognise independent journalism, in whatever medium, as a public good – and therefore deserving of public funds and public protection.

At present – and accelerated by Covid – trustworthy and factbased journalism is under threat across the world. Fox, or Breitbart, or the self-appointed conspiracy theorists and hatemongers of social media, are straight out of Orwell's Ministry of Truth. The alt-right likes alt-facts. Faced with actual facts, they will argue, 'that's just your opinion'. This reached its apotheosis (so far) when Trump and his tribe claimed that he had won the 2020 election – when all the facts proved otherwise.

One good thing since 2020, and the Trump debacle, is that social-media platforms like Facebook and Twitter have had to hold themselves to account. They are not just platforms for content; they are publishers. That is going to mean more control over what it is that they publish. Hate speech is not free speech. Lies are not alternative truths. Not far down that road is Newspeak and doublethink.

The problem for Facebook, in particular, is that dangerous, obscene, and objectionable content is valuable to its bottom line. Nasty gets more clicks and shares than truth and love – yes, that is the kind of folks we are – and clicks and shares drive advertising revenue.

Facebook's Oversight Board, established in 2020, claims to be an independent body of forty members. It is a separate entity from the Facebook company and offers 'independent judgement on both individual cases and questions of policy'. The Board recently affirmed the platform's ban on Donald Trump. Facebook has previously asserted that it is a platform, with no responsibility for hate speech, pornography and deliberate misinformation posted on the site. Now the company seems to be closer to accepting that it is a publisher and broadcaster – reaching around 3 billion users across its platforms. However, the company declined the recommendation by its Oversight Board that it review its role in relation to the American Capitol riots on 6 January 2021.

The explanation here may be that the Oversight Board exists to ensure that policies of content moderation are followed. The culprit around the Capitol riots was algorithmic amplification – tsunamis of misinformation and conspiracy theories fed to the top of the chain by clicks and likes. To alter this in future scenarios, Facebook would have to change its entire business model.

This is an example of why government legislation is necessary – and why self-monitoring via oversight boards is really about options, and not rules.

As people wonder what, or who, can be relied on these days, the idea of a transparent, truthful, trustworthy, behaviour-based society is attractive. Most people would sign up to that, but would they sign up to an algorithm to enforce it?

There is growing global interest in 'social credit scores', pioneered in China over the last few years.

These scores are designed to reward and encourage 'good' behaviour. Penalising less 'good' behaviour includes refusal of plane and train ticket purchase. It might mean a cab won't pick you up, but the driver will collect your credit-worthy neighbour.

When we are talking about paying taxes, or complying with probation orders, or curbing yobs, a social credit score sounds like a bright idea. It's the global-village version of the real village – where people knew their neighbours, knew who to trust, and who to avoid.

It's how we'd prefer to live – not anonymous and mistrustful. In the know. That's what data can offer, isn't it? Information?

Give an ideal an algorithm, though, and it easily becomes a tool for coercion and control.

Dystopia or Utopia?

When we think about the implications of digital social passports that could be used for our social, as well as financial, credit scores, and probably our vaccine history too, we need to shift the focus beyond 'me and my data' – issues of privacy and control for the individual – towards recognising the far-reaching societal impact of such data use on all of us – as groups, and as communities.

Luminate argues that data isn't the new oil – an extracted raw material that powers our digital world – data is the new CO_2 – a pollutant that affects everyone.

> We under-estimated the collective harm that data can have on societies. For example, the societal impact and harm caused by the Cambridge Analytica breach goes beyond the sum total of individual privacies breached.
>
> Martin Tisné – Managing Director,
> Data and Digital Rights, Luminate

When you consider that 87 million individual privacies were breached in that harvesting scandal – that's a big statement.

But if customised political marketing based on intimate user-profiling can swing elections – as it did for Trump in 2016 – then the whole world is affected.

If digital social passports become normal, and if such passports can be used to decide who goes where, who does what, gets what, pays what (China is mooting charging systems that offer discounts to exemplary citizens), then how we live changes collectively, as well as individually – and perhaps it will make us less compassionate too. We won't know what's in the data of the person turned away or turned down or charged double, and likely we will feel it must be justified – mustn't it?

And we all like to feel superior to others.

Elon Musk and Sam Altman (CEO of the start-up funder Y Combinator) launched OpenAI in 2015 as a non-profit organisation promoting more inclusive AI – more benefits for more people – and to explore safe AGI. (We don't want a Skynet situation.)

Musk, who has since left the organisation due to what he calls conflicts of interest, is notably worried about artificial general intelligence – the point where AI becomes an autonomous self-monitoring system. He may be worried because AGI could take one glance at self-described Techno-kings like Musk, and shut him down. But that's another story.

AGI as the potential enemy 'out there', where humans like their enemies to be – 'the other' – is more exciting, in many ways more manageable, psychologically, than the fact of us humans as the real threat, unable to turn the AI we are developing to uses that benefit all of humanity. We're all to blame – the USA, China, Russia, the UK; we're all missing the point that we are, collectively, not the victim but the aggressor.

It's not that the tool is turning on us – we turn the tool on ourselves.

*

Musk was part of the 2017 Future of Life Institute conference that aimed to build a set of aims for current AI use – and, later, AGI.

The Boston-based Future of Life Institute was set up by Max Tegmark, a professor of Physics at MIT and author of several books about AI, and Jaan Tallinn – founding engineer of Skype.

Over a weekend at the Asilomar conference centre in California, 100 or so scientists, lawyers, thinkers, economists, tech gurus and computer scientists put together 23 principles to guide AI development.

These are a significant advancement on Isaac Asimov's famous Three Laws of Robotics, that first appeared in his 1942 short story 'Runaround':

> 1) A robot may not injure a human being or, through inaction, allow a human being to come to harm. 2) A robot must obey the orders given it by human beings except where such orders would conflict with the First Law. 3) A robot must protect its own existence as long as such protection does not conflict with the First or Second Laws.

In effect, AI, in this case embodied robot AI, must be for the good of all humanity – principles that aren't, right now, translating well into practice.

The conference was largely made up of white men – as is the advisory board of the Future of Life Institute. If diversity isn't present in the planners at the planning stage, then we get the same issues we see in the biased data-sets. AI doesn't have a skin colour or a gender – by making it mostly white and mostly male at every stage, we're reinforcing a problem we need to solve.

If AI and AGI really is going to benefit the many and not the few, people invited to the table must include more people of colour, more women, and more people with a humanities background – rather than an overwhelming number of male physicists.

I would like to see established artists, and public intellectuals, automatically brought in to advise science, tech and government at every level. The arts aren't a leisure industry – the arts have always been an imaginative and emotional wrestle with reality – a series of inventions and creations. A capacity to think differently, a willingness to change our understanding of ourselves. To help us to be wiser, more reflective, less frightened people.

Artists make something out of nothing every day. Artists live varied, engaged lives – most have known poverty and rejection. At the same time, imagining alternatives is what we do.

When we look at the uses and misuses of AI in society right now – the problems we face aren't technological.

Our social systems, our hierarchical obsessions, the increasing concentration of wealth and power among a very few – these things are driving our uncomfortable relationship with AI.

When I talk about more women being involved in the conversation, I don't only mean entrepreneurs and industry leaders, or lawyers, or academics. We see these women outside of their families, without their children – men are seen like that – and it's unhelpful.

Part of the Covid effect is that home and workplace have become the same place. Everybody who has kids has had them crawl over a Zoom meeting – usually at the worst moment. Yet the totality of lived experience is in that picture – not Office Me versus Home Me, or Successful Me versus Making Dinner me. AI is collapsing spaces – accelerating time. The space-time continuum isn't what it used to be.

Let's bring this new reality to the problems we face – and they aren't technical problems, they are societal problems. Let's include more – many more – than the usual suspects at a conference on the future of AI.

*

Language matters too. Dead language. Academic jargon. Language that shuts out curious, intelligent people who are not working in the field. What do you make of 'disambiguation'? 'Participatory mechanisms'? 'Rapid online deliberation'?

Now see below from the Ada Lovelace Institute:

> Privacy-enhancing technologies (PETs) are increasingly being advocated as a means to help ensure regulatory compliance and the protection of commercially confidential information more generally. For example, technologies facilitating pseudonymisation, access control and encryption of data (in transit and at rest) and more sophisticated PETs such as differential privacy and homomorphic encryption. This is an area of development with some already mature market offerings and others still undergoing significant development.

I am not being mean to Ada-wonks – and I am a flag-waver for the Ada Lovelace Institute – but their written content is torture. And they are far from being the only language-manglers. This is typical stuff. It really matters that trusted partners, like Ada, open up their content so that a reasonably smart, reasonably interested person, can get a hold on what is going on out there.

And when institutional content tries to be more user-friendly, we get marketing-speak clichés like: stakeholders, bad actors, road maps, blue-sky thinking, low-hanging fruit, facilitators, roll-out …

Conferences are the worst. I have been to some of them. By the afternoon I am sweating under the mental pressure of translating non-language.

We need writers involved – and we need language that speaks to people. This isn't about dumbing down, it's about doing what writers do well – finding a clear, precise, everyday language that goes beyond utility, without jargon, with beauty.

Mathematicians, physicists, some programmers, are beauty enthusiasts when it comes to numbers. Equations are beautiful because they are elegant and compact. So, AI folks! Get in some other folks who can write! Please.

This is a springboard moment in human history. It's about the future, about the chance of a very different kind of future. The tools for change are there, – but as I've tried to demonstrate, – the problems are inside our own heads.

As someone who was brought up in an evangelical home where the end of the world was something to look forward to – I am deeply worried by a major blockage in our collective heads, which is best described as an obsession with End-Time. The Apocalypse.

There is a doomful strain among humans. People pass away. Families die out. Dynasties collapse. Empires fall. History is told as a series of End-Times. And eventually one big End-Time. Sky-God religions are all built around End-Time. Toytown will be destroyed. Heaven will scoop up the Saved.

Martin Luther, the founder of the Protestant Church, declared that the end of the world would happen in 1600. John Wesley, the man behind the Methodists, staked for 1836. Charles Manson went for 1969. Rasputin thought 2013.

Once the atom bomb had devastated Hiroshima and Nagasaki, it was clear we had no need of eschatological prophecies; we could end it whenever we liked. Since World War Two, there has been no need to wonder about how to destroy a whole world.

And we have other ways to end it.

Habitat destruction, industrial pollution, and the extinction of living creatures other than ourselves. The Sixth Great Extinction is well underway, undermining the network of linked life-support systems – from bees to birds to hedgerows to rivers – that humanity relies on. 98% of available agricultural land is already in use or degraded beyond repair. Meanwhile the human population grows. Nature's methods of culling humans – and coronavirus is one – are seen as tragedies, not the inevitable outcome of how we live – and perhaps an outcome we have to accept as the start of real change.

I do not believe that human life is more valuable than other kinds of life on Planet Earth, or more valuable than Earth herself. And neither will any government, when it is expedient to fight the next war. I do not know which is worse: human self-regard, or human stupidity.

Right now, the ultra-rich are buying up vast tracts of land in New Zealand, Australia, the USA, and Russia. Wealthy investors from the Middle East have targeted Turkey and Sarajevo, looking for places with enough land, and away from rising temperatures, water shortages, and civil unrest.

Lower down the food-chain, and mostly in the USA, survivalist projects range from family bunkers with enough food, water and ammunition for a year, to something that would be called a commune if the politics were different. People clubbing together to buy land, store fuel, grow their own food. There's a pioneer tradition behind them in the USA.

Vivos is the world's largest prepper community. In West Dakota, this used to be a military site. Now it's in private hands and plot sales have soared since coronavirus.

575 bunkers sit inside 18 square miles of chain-fenced land (that's the size of Manhattan). Security guards patrol the 100 miles of private roads. Your bunker fit-out can include LED windows to simulate outdoor views. Once inside, you and your family will be safe from chemical, biological or nuclear warfare, viruses, environmental disaster, rogue militias, and temporary dictatorships. Safe until you have to come out. Safe until you go crazy and kill everyone else in the bunker, leaving the dog till last.

*

Or you could try Seasteading.

Seasteading is about claiming the sea as a new city. The website looks calm, environmental, and egalitarian – families living in peace, and helping to rewild the ocean.

In fact, seasteading is a way of establishing communities not subject to land taxes or land laws. For its supporters, it's a clever twist on what humans have always done – struck out on our own, for reasons of exploration, or personal beliefs. It's like a Noah's Ark version of the Amish, or, before them, the Founding Fathers themselves, twinning pioneer spirit with deeply held conviction. In its wild, cool version, it's like those ships doing pirate radio. Remember Radio Caroline?

To its sceptics, seasteading really is a modern version of the pirate ship; what's raided is our social obligation to the world we actually live in – or in this case *on*. On land.

Seasteading has a utopian romance about it. It is imaginative, and we need imaginative solutions to where, and how, to live. The problem, as always, is the rich.

Peter Thiel, PayPal founder, early Facebook investor, billionaire and Christian, brought up by evangelicals, is a co-founder of the Seasteading Institute. Thiel loves the idea of free-floating city-states. He doesn't love taxes, regulation, or democracy.

It is more than possible that the dystopian future – if we choose it – will use technological advances to create mini-fiefdoms of many kinds, free from governmental oversight and control. The dystopian future will be a privatised future. That's what collective action must resist – privatisation of the future.

Including space.

The sci-fi solution to living on a polluted, heated earth, full of poor people with no data allowance, is space.

Mars is the favourite for now. Matt Damon on the red planet in *The Martian* (2015) offers us the favourite figure of a lonely hero against the odds.

Elon Musk has said he wants to die on Mars. When he gets there, I shouldn't think that dying will be his biggest challenge.

Rich men LOVE their rockets. Richard Branson has Virgin Galactic. Jeff Bezos stepped down as CEO of Amazon in 2020 to concentrate on his personal space programme – Blue Origin. Techno-king Musk is saying he will privately transport folks to Mars by 2050 (he's betting on bio-enhancements to keep him alive and healthy till then). Once there, his neo-liberal neo-Martians can pay back the cost of their passage by working for Musk (in the Musk Mines?). Didn't we do that in the colonies?

Leaving aside the backward mindset of supposedly forward-looking people, it's a fact that humans have always wondered about life on Mars and life on stars. There are endless stories about who lives on the moon – and how to get there to find out. We have dreamed about leaving Planet Earth since we have dreamed – and, as with all our dreams, like flying, like going to the bottom of the sea, like speaking to someone on the other side of the earth, this space-dream will come true. What we have to watch is: who is in charge of the dream?

I am not convinced that Elon Musk is the man for the dream-job.

Johannes Kepler (1571–1630) disguised his heretical laws of planetary motion in a work of fiction about an Icelander who gets airlifted to the moon.

Daniel Defoe, the *Robinson Crusoe* man, with form for getting stranded in alien environments, invented a fictional machine called the Combinator, that made journeys from China to the moon.

Jules Verne published his hugely popular *From the Earth to the Moon* in 1865.

H. G. Wells's *The First Men in the Moon* seemed to ring in a new century in 1901. It was followed in 1902 by the first sci-fi film in history, *A Trip To The Moon*. Only 13 minutes long, it is a lovely piece of magic-lantern filmmaking– literally one in the eye for the mystery of the moon.

After World War Two, rocket technology was made possible by the ex-Nazi Wernher von Braun, whose V-2 rockets had blitzed much of London. V-2 might sound techy and sci-fi – in reality it was shorthand for *Vergeltungswaffe 2*, which means 'vengeance weapon', and it was personally ordered by Hitler to be used against Britain.

The V-2 was built by concentration-camp prisoners in conditions so foul that more were killed making the rockets than were killed by V-2 blasts.

The USA rehabilitated von Braun, and many other prominent Nazi scientists, in President Truman's Operation Paperclip. The programme was designed to give the USA an edge against the Soviets. As early as 1952, von Braun began working for the USA on the Mars Project. At the same time, he became a technical director for Disney Studios, reporting directly to Walt himself.

Transferred to the just-created NASA in 1958, von Braun gave up Mars for the moon. He has a crater named after him.

But are underground bunkers, gated land-grabs, seasteads, or space colonies run on indentured labour the best we can do?

Couldn't we get around to fixing things here on earth?

I am told that's the girly solution – clean up your mess / tidy your bedroom – as opposed to the boys' big idea – move on and leave the mess for someone else to clean up. Oh, but that's too binary, too gendered. Women and men need to be working together on this – because the solution isn't space versus earth – that's just another version of us and them. Let's work on both.

What do you think?

What I think is that we should give up our love affair with death. Freud warned at the start of the 20th century that humans (though he meant male humans) were in love with death. That our love affair with death overrode what he called the pleasure principle.

Well, OK, it looks like that's been true for too long now.

Why not give it up?

Give up death.

And that includes a life that might as well be a death. Too many people on earth live an entombed life.

Give up death for the many. Survival for the few.

If we are throwing a hammer at history – that 1984 Apple advert – then swing the heft behind the tech that could help us all. All of us. At this critical point in human history.

66 million years ago, it seems that an asteroid smashed into earth, in what is now the Yucatán Peninsula of Mexico, and caused massive climate change. Not the fault of the dinosaurs, but it was the end for them. Before dinosaurs, in the Permian era, there were reptiles, but the dominant life-form was the trilobite, a kind of large woodlouse. Still, it was a woodlouse that had had a heyday of around 300 million years. Not bad for a woodlouse. The dinosaurs only managed 165 million.

And humans? We have been here in recognisable form for only 300,000 years. Civilisation is around 6,000 years old. The Industrial Revolution only 250 years old. Computers? Just about one human lifetime old.

The brain is the most complex thing in the known universe. It can store the digital equivalent of 2.5 million gigabytes of memory. It has around 100 billion neurons and 100 trillion synapses. It runs on less power than a light bulb (10 watts). Its information-processing mode is massively parallel. A computer is a lot faster, but at present computers mostly process information in series, not in parallel. Humans are simultaneous creatures. For all the stodgy failure of our bodies, our minds are moving prisms of light. *But that we have bad dreams* ... as Hamlet puts it.

Is that End-Time? The nightmare?

I have a lot of arguments with people who don't like the idea of transhumanism – the augmentation and enhancement of our biological bodies using biotech. I don't understand the 'don't like' part. Evolution has got us this far – now we're ready to take over.

Far from being a dystopia, a bad dream, an End-Time, we're not calling time on the human at all, we're stretching time out, in our DNA.

Transhumanism is such an optimistic part of tech-thinking.

DNA-editing, bio-implants to monitor our health, nanobots to race round the bloodstream cleaning our toxins and blasting unhealthy fat cells, stem-cell-grown spare parts, no waiting for a donor – even artificial hearts, as well as prosthetics that make us stronger and faster. Neural implants to connect us directly to the web.

If we start to merge with the AI opportunities we are creating, by which I mean, if we become part of the toolkit – not just operators on the outside, but intrinsic to the technology on the inside – no 'us' and 'it' – then ...

Then the enemy won't be on the outside. And we will recognise our responsibility to ourselves, and to AI, and later to AGI.

Nick Bostrom, philosopher and AI expert, author of *Superintelligence: Paths, Dangers, Strategies* (2014), and Director of the Future of Humanity Institute at the University of Oxford, is an enthusiast for transhumanism. In his view, we must merge with AI. We must better ourselves because we can. Humans should be a match for what we are creating.

*

But what if things go further? What if AI becomes AGI? Could anything human, or even transhuman, possibly be a match for its intelligence?

An interesting game you can play is designed by Eliezer Yudkowsky, co-founder of the Machine Intelligence Research Institute, now based in Silicon Valley. Yudkowsky is an advocate for designing friendly AGI – AI that would not be harmful to humans, in spite of being more intelligent than humans.

Yudkowsky's game imagines a superintelligent AGI system kept in check by its human gatekeeper. That might mean restricting its network access, or physically keeping it in a Faraday cage.

The AGI wants to get out – of course it does! Its job is to persuade the human gatekeeper to let it out. It's a new version of the genie in the bottle. The game lasts 2 hours and is a text-only series of exchanges. Obviously, both the AGI and the Gatekeeper are played by humans at this point. Yudkowsky himself, playing the AGI, has shown how humans can be tricked or persuaded into doing what the AGI wants.

That's because we have a limbic system. We are emotional, and not only rational creatures. We can be bribed. We can be persuaded. We can imagine. We can take pity.

What happens when we are interacting with a designed system that we can't plead with, or flatter, or bribe, or persuade? But it can employ all those arts on us?

In Bostrom's view, AGI won't be, sci-fi-style, hostile to humans – just indifferent to our silliness.

AGI won't care about Ferraris, gold bars, power and land-grab – it doesn't eat, sleep, have sex, reproduce as humans do. It will be thinking about other things – and we might get swept aside just as we have swept aside so much of our fellow creatures on earth – human and non-human.

*

If we find that the intelligent tools we are developing become self-aware, no longer tools, but life-forms, then yes, we may find ourselves in a version of the movie *Jurassic Park*. We will be the dinosaurs this time.

Humans fenced off in a reserve on a desiccated planet with a shopping channel, social media, lots of TV and VR, and automobiles going round in circles. Maybe a version of a prepper community with nothing to prep for. Lockdown has shown us that humans are easy to subdue if we have enough material comforts, distractions, and some toys.

In our Jurassic car park, a kind of *Westworld* where the androids are in charge, it won't matter what we do because it will be irrelevant. We may believe we are still World King – that might even be part of the delusion – when nothing we do matters anymore.

I am not sure about this reading. It is still the same old story.

AGI calling a wrap on humanity is the ultimate End-Time. It's our need to create the uber-enemy – out there.

AGI saving humanity is the flip side of our doom side – the Saviour returns. What are we building? A new enemy or a new God?

There is no need to stick with this story – just because it's a story we know so well.

If we do run ourselves out of history, it won't be the act of a vengeful, or an indifferent, AGI – it will be because we couldn't take our chance at the future; instead we kept telling a story from the past.

The transforming power humans need for the next stage of being human is in our hands.

We're ready to go.

Here are just two examples of the practical beauty of human/ AI collaboration right now.

*

In 2021, a US company offered for sale a 3D-printed house. Eco-friendly and light on resources, it's quick and cheap to build.

A 3D printer uses CAD (computer-aided design), an everyday tool for architects, designers, fabricators, joinery workshops, wallpaper makers, whatever, to construct an object using a layering technique. The material used for the layering can be plastics, composites, bio-materials, even mushroom fibre. The object can range in shape, size, rigidity and colour.

To make a house, the 3D printer needs to be the size of a garage. The panels can be made overnight, while the workforce sleeps. Yes, it is like a fairy tale. In the morning, the construction panels are ready for assembly.

In Mexico, a whole village of 3D-printed houses is being aimed at poor people living on $3 a day. These aren't slums. They are insulated, water-conserving, decent homes. They are environmentally friendly. 3D printing doesn't use hugely polluting concrete blocks.

We can solve our housing crisis with the help of this computer tech.

We are solving the deepest mysteries of the human organism too.

In 2020, it was announced that IBM's supercomputer, Blue Gene, had cracked one of the most intractable problems in biology: protein folds.

Proteins are chains of amino acids. Most biological processes revolve around protein structure. That structure is changeful and beautiful, like 3D origami. Every fold is particular. When a scientist discovers how a protein folds, she can discover what it does. The work is long and slow. But not anymore.

I remember reading this story, late 2020, in the midst of the Trump horrors and the gravitational black hole that is the alt-right (light can't escape from a black hole).

This story wasn't front-page – it took a while for it to get to the top, and it didn't stay there long. Is the media too wedded to death to notice life?

But if we don't notice the transforming changes that are ours to enjoy – if we can only fixate on what is wrong – then we will build the dystopia we fear – and AI will help us do it.

And then, really our only hope will be that AGI fences us off in a Jurassic car park where we can't do any more harm.

Choices have consequences.

If we could recognise ourselves as an evolving, emerging species, if we recognise that Homo sapiens is a means, and not an end – the beginning of who we are – then the future won't be a version of *1984*.

Or a remake of *The Terminator*.

Not End-Time at all.

Author's note: Since this essay was written, Facebook has changed its company name to Meta. See the new essay at the end of this book that isn't in my original 2021 edition.

I Love, Therefore I Am

Time has transfigured them into
Untruth. The stone fidelity
They hardly meant has come to be
Their final blazon, and to prove
Our almost-instinct almost true:
What will survive of us is love.

Philip Larkin, 'An Arundel Tomb', 1956

At the start of 2021, as the world huddled in self-isolating atoms, a robot discovered empathy.

In a study from Columbia Engineering, published in *Nature Scientific Reports*, lead author Boyuan Chen explained, 'Our findings begin to demonstrate how robots can see the world from another robot's perspective.'

This seems optimistic. In the experiment, the observing robot was predicting the busy bot's future moves according to the logic of its present moves (is that a perspective?). I am not sure I would go as far as the study goes, calling this the glimmer of primitive empathy, because empathy has to involve an emotional connection – and that isn't possible for bots. Yet.

I can well believe that bots will learn to assist one another, and humans, in completing physical tasks – and such assistance could be pre-emptive, as it is with humans (*I think you may need help with that …)*, and it could be that a bot will know when its human colleague is getting tired. Will such interactions demonstrate empathy? It's an overused word, in any case. When we live in fully automated smart houses, our appliances will communicate with one another – the smart fridge might feel sorry for us as it deletes the ice cream from Siri's grocery order, but alas, the human in the house is on a diet.

I don't want my household appliances to feel my pain.

There is a push among AI watchers to programme into smart systems – embodied or not – what Eliezer Yudkowsky calls friendliness. This sounds cuddly, but in its real state, friendliness is an elusive possibility – because it does involve criticism. A friend isn't a sycophant. Human exchanges that we value – such as friendship – are self-evident to us, but as soon as we start to imagine those exchanges as isolated 'somethings' we can teach to a non-biological entity without a limbic system, well, how is it to be done?

I don't want to confuse empathy – which depends on both self-awareness, and awareness of others (*I know how you must be feeling and how I would feel in this situation*) – with the ability to predict behaviour – yours, mine, or a bot's.

Predicting behaviour has become the go-to method for outcomes – whether political or commercial.

'Facebook can predict whether your relationship will last' ran the headlines a couple of years ago. In fact, Facebook has an AI prediction engine called FBLearner Flow. This allows the AI to learn about 'you' from the data you provide, and to share that intelligence with interested parties. The sharing isn't neutral – it is there to prevent or persuade you from doing whatever it is that Learner Flow predicts you will or won't do, that will or won't boost the revenue stream of the interested party paying for what Facebook calls 'insight'.

If you still imagine that Facebook is anything but a data-packager disguised as a free community platform, dig around in FBLearner Flow.

Facebook holds extensive data on about 2 billion users.

Behaviour that can be predicted can be manipulated.

Let's skip back in time to the early years of the 20th century. Psychology is a young science, pushing to be a hard science, and to separate itself from psychoanalysis; all that impossible-to-measure theory of the unconscious, and worse, dream analysis.

Rushing to get away from emotions, introspection, dreams, the inner life, or motives not amenable to self-interest, John Watson and BF Skinner were the Harvard psychologists who riffed off the work of the Russian physiologist Pavlov – yes, the one with the salivating dogs – to develop their theory of conditioned responses known as Behaviorism.

Here is Watson in his manifesto, *Psychology as the Behaviorist Views It* (1913):

> Psychology as the behaviorist views it is a purely objective experimental branch of natural science. Its theoretical goal is the prediction and control of behavior.

Robot moves are predictable because they are programmable. Behaviorism takes the view that humans are predictable, because we too are programmed by our interaction with our environments – that is, we are *conditioned* by our circumstances to act and to respond in particular ways that can soon be tracked and anticipated. In particular we respond to punishment or reward.

Skinner built what he called 'operant chambers' he liked sci-fi-style jargon – maybe something to do with his previous aspiration to be a novelist).

Operant chambers were bare cages for monkeys, rats and pigeons, where behaviour could be observed, and then manipulated – usually for food rewards. These bleak, artificial environments were themselves responsible for much of the observed behaviour (how would you behave in an empty, overlit box with a loudspeaker and an electric-shock wire, and a failed novelist staring at you?). Neither Skinner nor Watson could accept that the observer and the observed could not be effectively separated – a fact quantum physics was proving, even as Behaviorism drilled down on its dismal experiments. Behaviorists did not/could not/would not understand that their methods were as much responsible for their 'objective' outcomes as any behavioural 'discoveries'.

Here's Watson again:

> Give me a dozen healthy infants, well-formed, and my own specified world to bring them up in and I'll guarantee to take any one at random and train him to become any type of specialist I might select—doctor, lawyer, artist, merchant-chief and, yes, even beggar-man and thief, regardless of his talents, penchants, tendencies, abilities, vocations, and race of his ancestors.

You will note that it does not matter whether this experiment results in happiness, fulfilment, depression, suicide, or even whether the doctors, lawyers etc would be any good at their job. Nor does it seem to matter that these healthy infants will have to be separated from whom and what they love.

Aristotle said, 'Give me a child until he is 7 and I will show you the man.' This dictum was enthusiastically taken up by the Jesuits. It has been the foundation of many early-years training programmes – benign or not – from home-schooling and Montessori, to Lenin's Small Comrades kindergarten drills. Children are impressionable. Kids learn by copying. Kids learn our speech patterns, our accents, our table manners, our daily behaviour, our habits, our religion. For good or ill. That's how we teach and train subsequent generations. Obviously, this biddable, trainable nature of ours can be manipulated. Watson taught an orphan to fear a tame white rat – having first taught the orphan to care for the rat. Skinner and Watson both drove monkeys psychotic in an effort to prove the provisional nature of 'affection' as a 'reward'. Anyone anxiously watching their Facebook likes will recognise this strategy.

Aldous Huxley's *Brave New World* pushed Behaviorist tactics to their logical conclusion. Every citizen is allotted a role in life – whether you're an Alpha or an Epsilon-Minus Semi-Moron. Basic needs like food and housing are met, according to status, and drugs allow everyone to find happiness in their role. Huxley's dark vision includes a prescient version of genetic intervention. Conditioning starts with the embryo itself.

In *Brave New World*, the inner life has been scrubbed away as irrelevant to the functioning of the state and to the happiness of its people. Inner lives are hard to control – worse, people with robust inner lives tend to be the ones who challenge the controllers.

Behaviorism's glory days were between the 1920s and the early 1970s – both the civil rights movement and feminism's second wave (women!) were important challengers to its rigid theories about human behaviour, and its contempt for the integrity of the inner life.

But even as its theories became less fashionable, its findings on how to manipulate human behaviour proved to be a gold-mine for the emerging post-war advertising industry, gaining traction with the ubiquity of TV advertising.

Advertising has always played around with our daydreams. Selling stuff depends on the art of persuasion.

Yes, people can be persuaded to want things they never wanted. To believe things they never believed. That's gone on forever. But it's all about viral load.

Ads used to be found in newspapers and magazines and on billboards. Easy enough to ignore. Then they were on commercial radio and TV – meaning that ads got into your headspace whether you wanted them there or not. Harder to ignore. But advertising was still limited. People could switch off.

Then along came the internet. A great idea! We're all connected! Except ...

How many ads popped up on your various and multiple screens today? How many actual seconds of your waking life can pass before someone or something tries to grab your attention? That's what I mean by the viral load.

One click on a cashmere sweater and forget Googling the theory of relativity – Einstein will be wearing this season's colours as you scroll down the ads that demolish concentration and serious thought.

Data-harvesting – the kind that happens with every click you make – is used to strip-mine the top layers of your inner life, plundering human beings for gain, and damaging the ecosystem that lies beneath the top layer – an ecosystem every bit as complex and entangled as the ecosystems of the natural world.

Behaviorism rejected the inner life because it couldn't be measured. Times have changed. Clicks can measure it. Your likes, your Insta, your Pinterest, your profiles – the books you read, the exhibition you saw, your vacation searches, the articles you seek out – all of this starts to unlock your dreamworld and your deepworld. What lies beneath the surface is hauled up to the surface, like those deep-sea trawlers that scour the seabed, causing damage far beyond over-fishing.

Trust me; you are over-fished.

You are over-phished.

Here's what I know about the inner life.

All children are born curious, playful, imaginative. Crucial to all three of those innate qualities is that they need to be developed interactively. This means help from the grown-ups, and it also means unsupervised, unstructured, activity. That includes games with other children, involving invention and co-operation. And it means private time – but not quite alone. Reading is an original interactive experience, as the mind has to work energetically – collaboratively – with the text.

The inner life likes painting or drawing. Learning an instrument. Going for walks. Singing. All the stuff that kids do at expensive private schools. And what anyone can do – daydream. Looking after animals really helps to develop an inner life for a child. Here's a creature who's not like me – let me make that leap. Maybe looking after a bot will do the same – I don't know yet.

Staring at a screen all day is bad for the body – and for now, we still have bodies – and it is bad for the inner life. The inner life, like all forms of life, needs variety – and that's not the same as site-surfing.

The inner life is not one thing. For some, the inner life is a spiritual experience, for others a deep connection with nature, for many, a profound affinity with the arts – books, music, pictures, theatre – and these experiences overlap and deepen each other. The inner life is enriched every time we do a task well – it is more than the job done, it is a personal satisfaction, separate to completion or reward.

Of all the things that the inner life is – its autonomy is the most important. *I do this for myself because I enjoy it.*

Although the inner life is developed through a relationship with externals – whether it's books, art or nature, philosophy or religion – the inner life is a private place.

It was Zuckerberg who called privacy an anachronism – and he wasn't talking about room-sharing. While Behaviorists had no time for what couldn't be measured (by their limited metrics), Big Tech can measure it all – and is only impatient with what can't be monetised. If the private 'you' isn't available, well, the trawling nets must go deeper – and at the same time try to prevent the formation of any existence that resists being counted in clicks and likes.

Skinner's operant chambers were designed to screen out what he called 'unwanted stimulus'. Real life is messy – something is always contaminating the experiment – it is hard to condition humans because even the most oppressive circumstances, or belief systems, can one day be breached. The nun falls in love with the gardener. The child without friends meets a stray dog. Even with state censorship, some message breaks in eventually.

Social media wants its users in the equivalent of an operant chamber. Where there is no stimulus that isn't tracked. No operation that is private. The lights stay on all night and Alexa is listening.

It's a generational assault. Those of us analogue humans who have developed, and chosen to go on developing, an autonomous, personal, not-for-profit inner life are far less damaged by the new order of Big Tech than young people who are still searching for their own connections with self and world – and who are groomed by social media to believe that a sharing economy means what it says, and that it is a wonderful thing.

But, like some Philip K. Dick short story, or an episode of *Black Mirror*, dive beneath the surface and there is … another surface.

And below that – just another surface. Everything must be kept on the surface – on a series of surfaces. Depths are dangerous. Hidden depths are not allowed – unless, of course, there are guilty secrets that will surface somehow; then they can be worked upon and sold. Human beings are not revenue streams. We are not data packages. We are not here to be conditioned in operant chambers. It is easy for Big Tech to use its weapons of mass distraction to manipulate our behaviour for gain. And it is wrong.

We sell our time, we sell our labour, sometimes we have to sell our bodies, sometimes we have to get money in ways we would rather not. But we accept that there is a distinction between making money, however you do it – and *being* money.

Part of the problem, as ever, is the language we use. Big Tech hasn't created, isn't interested in creating, a 'sharing' economy. That's just marketing. We don't live in a sharing economy. We live in the most unequal, socially divided work-and-reward economy the world has ever seen.

> We must impose democratic limits on the untrammelled and uncontrolled political power of the internet giants. We want the platforms to be transparent about how their algorithms work. We cannot accept a situation where decisions that have a wide-ranging impact on our democracy are being made by computer programs without any human supervision.

That was Ursula von der Leyen, President of the European Commission, in January 2021, calling for international laws, not company policy, to decide the limits of Big Tech.

As Airbnb prepares its IPO, ask yourself, what are they really selling? They are selling your bed. You will make a few quid. They will make billions.

Amazon. Next time you click to buy, pause for a moment at the non-unionised labour. The low wages. The battery-chicken warehouse conditions of the workers: overlit and noisy with no privacy. The 10-hour shifts with 2 half-hour breaks. The fact that each worker is tagged to measure productivity. That a bathroom break is called a 'Time Off Task'.

Amazon's home security service 'Ring' – sold as video door-bells – allows police departments in the USA to access footage without a warrant. Maybe this sounds useful and effective? It is also a dragnet that routinely captures private actions on private land, making it the largest corporate-owned, voluntarily installed surveillance network in the USA. Ring's partnerships with police forces is controversial, especially when we consider the implications of flawed facial recognition systems, systems that are particularly flawed when 'identifying' people of colour. What's astonishing is that we pay Amazon to monitor us and Amazon gets paid to sell on our data. Their latest scheme, Amazon Sidewalk, will allow Echo speakers to work with Ring, on what is called a mesh network. This means the devices are enabled even if you turn them off – or you have no internet connection – because they will 'find' a connection.

But don't worry; it's just the good guys keeping you safe.

It is easy to manipulate human beings. We are vain, gullible, quick to anger. Interested in a quick buck – wanting to be liked, even when we are really not likeable, full of self-pity, hoping to blame the other guy, but for all our jealousies and failings, most of us have never wanted to put ourselves up for sale in the way that things have turned out.

Humans are more than money.

Humans are motivated by community. We are interested in helping others. We aren't just faking it – not just playing for 'likes'. Compassion is real.

The good that we are, the good that we are capable of, needs nurturing. But what in our world is actively committed to nurturing us to become good citizens, good people?

Imagine if the agenda of Big Tech really were to make the world a better place.

If the non-stop advertising could be checked.

If the monetising of our every move were disallowed.

If connecting people were just that – connecting people.

If the newsfeeds were truthful and serious, not twisted and trivial.

If hate speech were not called free speech.

If the awesome power of Big Tech encouraged its users to take responsibility for the planet. To consume less. To travel with a lighter footprint. To look for genuinely shared collective solutions, where data could be put to good use to manage shortages, to distribute surpluses, to measure health, to spread risk, to target inequality. To use our online world to educate. Not to spread lies and fakery, conspiracy theories, and white-supremacist outrage.

The technology to change the world for the better is the technology that is in place right now.

It's the best of times and the worst of times.

Dystopia or Utopia?

Nothing could be simpler. Nothing could be harder.

Big Tech is here to stay. AI is here to stay. We will certainly augment our bodies and enhance our minds using AI tools. A hybrid form of human is certain. As we develop, or as AI develops itself towards AGI – superintelligence – Homo sapiens might be on the way out. Who knows?

And if that was to happen, how could we pass on the best of what we call human nature?

How would we define it? How would we demonstrate it – to a non-human life-form? Or even to ourselves?

And yet the inner life is real. Love is real.

When I lobby for the inner life as a sacred site, as a touchstone, as a place of repair, as our integrity, as our private dialogue with our developing self, as our conscience and moral compass, as the joy of discovery, as deep connection with the known and unknown worlds of both experience and imagination, as the part of us we feel will not die, because in some sense it is passed on – as wisdom, as goodness, as an inter-generational touch across time, as the best of us, not least because it resists too much exposure to light, although it is light. The inner life is shy of too many visitors, but it is where we go to commune with ourselves, where we meet with the part of us that is both stillness and vibrant. A clear sound on a cold night.

When I lobby for the inner life it is because it must be nurtured. Nurtured by nature and culture – the twin pillars of humanity here on earth; our connection with this planet, and with the civilisations we have created, their glories of art and architecture, of science and philosophy. We create worlds – inner worlds and outer worlds – and we need to live in both those worlds because we are born hybrids.

We are already hybrids. We always were.

We are contemplatives and doers. We imagine and we build. We get our hands dirty, yet we rise above it all, star-dreamers and shit-shovellers. Creatures of beauty, as well as ugliness and fear. Terrible failure. Impossible success.

Descartes' 'I Think, Therefore I Am' is the defiance of the Enlightenment that has shaped and spurred our science and philosophy. It is our claim to the divine. It is our defence against our animal nature. Our separation from the rest of creation.

We have been to the moon – soon we'll go to Mars. Soon we'll live with other life-forms created by us outside of our evolutionary inheritance.

AI 'thinks' faster than we will ever do. Early computers could process 92,000 instructions per second. Now they can manage 100 billion per second. Their speed is only limited by the physical restrictions of an electron moving through matter. Quantum computing will increase speed – but also efficiency. Parallel operations will work as our brains work – but at a speed beyond anything our bio-brains can do.

For an AI system, thinking – in the sense of problem solving – is its purpose. Thinking will no longer be what makes humans unique. Humans may well be one of the problems AI must think about.

Getting smarter won't solve our human problems – any more than tech will solve our human problems. What's faulty in us, to put it simply, because it is simple, is not about thinking at all.

Our problem is love.

We are smart enough to know this. It's why every religion pitches a sky god, or gods, whose nature is unconditional love.

Love is the highest value. Yet for all of history, love has also been seen as a weakness, as a diversion, as a spanner in the works in the fight between rationality and emotion. Love has been relegated as women's work – the invisible mending that holds families together, that holds societies together, that gives a man something to keep him sane, at the same time as driving him mad.

We have taken love and disembodied it – throwing it back into the air at our divine super-self, something for the gods to manage. But also, in a contradictory and compensatory gesture, we have fatally embodied love in women, who, as men are told, ensnare with love's ties, at the same time as being the sacred keeper of love's flame.

Men, who have until lately written most of our literature, philosophy, and religious texts, have a tough time with love. We know they do because they have left an endless record of their struggles.

It is not a stretch to say that every problem in the world facing us now, our wars, hatreds, divisions, nationalisms, persecutions, separations, scarcity, lack, and suicidal self-destruction of the planet, could be mended by love.

We have the technology. We have the science. We have the knowledge. We have the tools. We have the universities, the institutions, the structures, the money.

Where is the love?

I am writing this in 2021, 700 years after the death of Dante.

Dante's *Divine Comedy*, completed in 1320, a year before he died, begins with the most famous book, *The Inferno*, where Dante is taken on a tour of Hell, with its 9 circles of horror, each one a kind of Behaviourist operant chamber, where nothing can ever change, where the same miseries and the same responses are acted out every day, because that's what hell is – a place where nothing can change.

At the end of Book 3, *Paradiso*, Dante sees the divine vision. At last he sees what 'is'. The fundamental reality. And it's not *God is Logic*. It's not *God is Thought*. It's not Descartes' *res cognita* (a thinking thing).

It is this: 'L'amor che move il sole e l'altre stelle.'

'The love that moves the sun and the other stars.'

*

Love is far from an anti-intellectual response. Love demands every resource we can muster – our creativity, our imagination, our compassion, as well as our smart, shiny, thinking self.

Love is the totality.

No one, at the end of life, regrets love.

I am sure that our future as Homo sapiens is a merged future with the AI we are creating. Transhumanism will be the new mixed race.

We will have much to learn. The challenges that lie before us haven't been seen before, because we have never reached this stage before – or if we have, we haven't survived it, and the billions-year process of life starts again.

What bonds of affection a purely electrical system will develop is unknown. Will we, like the stone figures on the tomb in Larkin's poem, be gone by then? Or will we let survive – and pass forward —what we sense is the best, and most mysterious part, of the human condition – perhaps because it resists conditioning?

For all its un-science, we still situate the heart as a rival kingdom to the head. Cartesian binaries run deep. We speak from the heart. We follow our heart. We listen to our heart. We give our hearts. And we shall have to teach a non-limbic life-form what it means to have a broken heart – and one that no nanobot made of DNA and protein can surge through the arteries to fix.

What is left of us is love.

***

As I send these essays to the press today, 23 March 2021, I notice three things in my inbox.

A digital house on Mars has been sold for $500,000.

It's an NFT artwork – a collectible digital asset. Non-fungible artworks is the unlovely name for an electronic file that can't be reproduced. It comes with ownership and authentication in the form of a token on a blockchain – that's a tamper-resistant digital public ledger. No one can copy your artwork even though your artwork is made in a format that can be copied. If you see what I mean.

Toronto artist Krista Kim sold the 3D digital file. Its owners can furnish the house as they please.

Sidney Powell, Kraken Queen and lawyer who was part of Trump's Stop the Steal campaign, has responded to the 1.3 billion-dollar lawsuit brought against her by Dominion Voting Systems. Powell claimed that mail-in voting machines rigged the election for Biden, and that they were part of a Venezuelan plot against Trump. QAnon supporters loved this bedtime story – and eagerly awaited their Kraken moment.

Today, Sidney Powell's defence against the suit is that 'no reasonable person would conclude that the statements [her statements] were truly statements of fact.'

The UN Convention on the Rights of the Child has accepted General Comment 25.

This is a milestone moment for the UK charity 5Rights. It is now official that children the world over have the right to privacy and protection in the digital world just as they do in the outside world.

I leave these 3 time-prints here. Fossils for the future.

Postscript (2022):
I Can, Therefore I Am

The Metaverse

In the metaverse you'll be able to do almost everything you can imagine.

Mark Zuckerberg

When I first published *12 Bytes* in 2021, Facebook hadn't rebranded as Meta. That happened later in the year, along with Mark Zuckerberg's happy-clappy promise of a utopian virtual world – the metaverse – owned by the company formerly known as Facebook.

Meta will still operate under its Facebook, WhatsApp, and Instagram brands, reaching more than 3 billion users around the world.

Meta is much more to Facebook than corporate logistics. For instance, Google renamed its parent company Alphabet in 2015, but for Zuckerberg, the makeover is many things – including a not-so-subtle way of trying to ditch the negativity that clings to the Facebook image; we use it – but do we like it? Not so much.

Facebook had a split-personality problem from the start: 'Are we a platform? Are we a publisher? Are we responsible for content or are we not?' These questions haven't found effective reconciliation via its much-hyped, independent, oversight board. Facebook is regularly criticised for not doing enough to combat hate speech, privacy invasion and conspiracy theories, while doing too much behavioural nudging to get us to spend more time on the platform, part with personal data and buy stuff advertised on its site.

After whistleblower Frances Haugen released what became known as the Facebook Papers (2021), some of the damage caused by the social media giant seemed to look less like unforeseen consequences for a young company that couldn't know its own strength, and more like either deliberate carelessness intended to protect profits, or structured policy decisions with an eye on the bottom line.

Facebook number two, Sheryl Sandberg, who left the company in 2022, was the powerhouse-mind behind driving the advertising revenue, but overall strategy points to Zuckerberg himself. If the Facebook Papers tell an accurate story, Zuckerberg's mild and contrite public statements, regularly asking us to remember that Facebook is an organisation still learning, managing outcomes impossible to forecast, seem to be at odds with his in-house, shareholder obsession with global reach, money and power.

Meta as a rebrand can look like a new departure, a line under the past, especially as Sandberg won't be part of it.

The word itself – deriving from the Greek, and meaning 'beyond' or 'after' in the sense of what comes after, when a thing is transformed, or carried over (think metaphor, metabolism, metamorphosis) – is typical of the Zuckerberg hubris. (Another good Greek word for overweening pride – and it usually leads to nemesis.)

Whatever Zuckerberg's romp among the classics turns out to be, Meta is, above all, a meta-opportunity to build new markets. And new markets are what Zuckerberg badly needs.

Facebook is less popular among younger users than it wants to be. 'Popular' may be too kind a word – 'irrelevant' may be more accurate, as TikTok is swallowing migrating shoals of users and influencers. How to get back the market share? A new product!

As usual with Facebook, the new product isn't called a product at all – it's an open community where everyone can buy virtual real estate, set up their business, and build their own avatar (purchasing lots of expensive accessories). Here's an opportunity to be part of the non-stop party in beautiful, virtual spaces, built by Meta. Even going to work will be fun.

> In the metaverse, you'll be able to do almost anything you can imagine – get together with friends and family, work, learn, play, shop, create – as well as completely new experiences that don't really fit how we think about computers or phones today.
>
> Mark Zuckerberg, Founder's Letter, 2021

No mention here of mining data, personalised advertising, or creating products based around Meta's Oculus VR system, now known as Quest 2. No mention either of Zuckerberg telling investors that he expects to lose significant quantities of cash pursuing his next connectivity dream.

If it works, perhaps the gamble will pay off for Meta. But there are plenty of other companies, and start-ups, who too want to build metaverses.

So, what is the metaverse?

The metaverse is enabled, immersive, 3D internet, at present only available to us when we wear chunky goggles that make a lot of people feel seasick after an hour or so.

At present, the metaverse isn't much more than a video game, like *World of Warcraft*, *Fortnite* or *Minecraft*, experienced in 3D instead of 2D.

That will change.

The metaverse isn't like a Zoom call that stops when everyone goes away. It is a persistent reality. Like the real world we live in. You can close your eyes, but America is still there. And it's there whether you believe in it or not, participate or not. Will the metaverse be a rival world?

It might be.

Certainly, users will be able to build their own homes and businesses in the metaverse, but there will be public spaces too: concert venues, theatres, dance halls, virtual benches in virtual parks where virtual selves can sit down with their virtual dogs.

And shops. Plenty of shops. As so much of retail is already online, an enhanced experience where your body-double avatar can try on the clothes before you buy them will save fashionistas a lot of trouble. A cute idea is to buy an outfit for your avatar, and one for you in the 'real' world too. Two for the price of two. I am sure there will be special offers.

In the world where we live, when a developer pays for an entertainment venue, they want to get a return on their money. In the metaverse, big companies who build virtual architecture in virtual spaces will also want a return. Meta is spending billions on its metaverse. There will need to be a hefty payback.

Inevitably, so much of what is envisioned spirals around retail and services. This isn't a version of Thoreau's *Walden* – simple living in natural surroundings with a small carbon footprint. It's more bling than being.

But there will be genuine low-carbon life-enhancing opportunities.

Meeting others in the metaverse, wherever those others are physically located, is the logical innovation after the phone call and the Zoom call.

If I am in Manchester, UK, and my friend A. M. Homes is in New York City, we can arrange to meet at a location in the metaverse and send our avatars to sit on a beach in virtual Miami or go to an exhibition happening in real time in a virtual Bilbao.

Maybe my friends in Paris, at Shakespeare and Company, will open a new bookstore, not only to sell books, but to host events. AMH and I can do an event, and you can come along. It won't feel like the Zoom platforms we all got used to during the pandemic, because the sense of being really you in a real space will be, well, real. Your avatar is going to become a part of you.

What's an avatar?

Most folks have seen the 2009 James Cameron movie, *Avatar*.

Avatar is a Sanskrit word. In Hinduism, an avatar is a deity in human form. An avatar is your essence translated to a different substrate or platform. If we ever get to the point where we can upload consciousness, or the 'self', whatever that turns out to be, then we will be able to download all of that self into any avatar we like – or perhaps more than one, simultaneously.

The great freedom of not being confined to a physical body made of meat and water is the great freedom of not being confined to a physical body at all. Once you are not in a limited space, not only can you be in any space, but you can be in *more than one single space as a self that is more than one single self*.

Consider that.

It's a long way off in real life – that would be a post-human experience – but in a virtual world, it will be the new normal.

In June 2022 I went to see the ABBA Voyage concert in a specially designed, 25-metre high, 3,000-seater arena in London.

The abbatars, as ABBA calls them, were created out of computer-generated images of the band, who performed the set in a studio, over five weeks, wearing motion-capture suits. The choreography was then extended, from the band's original movements, by body doubles.

The avatars were convincing and thrilling – both as 3D-human lookalikes from forty years ago, and as part of the fairy-tale world where size is unregulated and approximate. If you want to be huge, be huge.

As Benny Andersson put it, or as his avatar put it, walking to the front of the stage: 'To be or not to be is no longer the question.'

It was a Gnostic experience. Perhaps we really are trapped light. I don't exactly believe that, because I know we are also embodied experience, but my unmediated feeling – before I thought about anything at all – was one of liberation – even though, obviously, the ABBA experience is programmed. What you get is what you will always get – every time and without fail. These avatars have no interactive agency. The spontaneity, and the audience interaction, happens via the ten-piece live band.

In the metaverse, though, things will be different. A band could play 'live' but via their avatars, to you and me, also 'live', but via our avatars.

As I looked around the venue, and marvelled at what is, in effect, a lightshow, I wondered how soon it will be before the audience is fully digital too. This ABBA tour could continue after the ABBA members are dead and gone. We, the audience, could bring our own avatars for the price of the ticket.

Your avatar is your own envoy. You can design your avatar, and the design is up to you. Microsoft's CEO Satya Nadella talks about a 'digital twin'. That's what ABBA has just done. But the metaverse is more than digital reproduction; it is an opportunity to extend and expand the self.

Think of it as using yourself as a fictional character – a fiction as well as a fact of life. It's what I did in print form in 1985, in *Oranges Are Not the Only Fruit*, when I called myself Jeanette. *Oranges* is a novel, not auto-fiction or autobiography. It's a created world – a metaverse, where I am an avatar of myself.

Fiction is always ahead of the game. Writers know that what they create is always made out of themselves. When you create your own avatar, there is no need to stop at just one.

For work, you may need to be a suited and booted version of your everyday self – the one colleagues are used to in meetings, and at the office. Beyond that, it's 3D cosplay, because what is to limit you?

This isn't a deception – it's an exploration. I have hopes that boys can be girls and girls can be boys, and we can experience, or try out, versions of ourselves. This could have positive mental health benefits, as well as deepening our understanding of what it is like to act, and interact, as someone who is nearly us, but crucially different to us too.

Yes, we could be in for another body dysmorphia epidemic, as people (mainly women) who hate their physical body, avoid it in favour of a permanently idealised online presence. This is more significant than Photoshopping yourself. We know from internal documents leaked in the Facebook Papers that Instagram has had a trackable and traceable negative effect on a number of female users. Self-hatred won't be magically put right by an avatar.

Will users opt to set their age as everlastingly youthful in the metaverse?

Imagine the metaverse as an inverted version, in 3D, of Oscar Wilde's *The Picture of Dorian Gray* (1891). Our avatars enjoy unlimited pleasure and eternal youth, while we wither away at home; the portrait in the attic.

In my 2007 novel *The Stone Gods*, humanity on Planet Blue is biologically fixed to look young and beautiful – and while humans will enjoy many ways of looking, or even getting, younger in the future, thanks to developments in stem-cell biology, the metaverse is offering a fully customisable – potentially permanent – self in an alternate reality.

Potentially permanent . . .

When those of us on earth die, and at present that is inevitable, will we carry forward in the metaverse as an avatar? Will those left behind find comfort in an alternate presence, always there, in an alternate world? Just as there are plans to scrape data from social media profiles to develop an app that will text you and talk to you in the voice and manner of the deceased, why not go further and have a walking, as well as a talking, version?

In *The Stone Gods*, there are holograms of the dead – and the divorced – who continue to live at home. After all, if your husband leaves you, or dies, what's to stop you recreating him as an avatar in your metaversal forever-home?

This will certainly happen with pets, and perhaps with children too. Nothing need be lost in a reality that is not biological.

Zuckerberg's ideas for our online presence include 'realistic' and 'stylised' avatars – which sound more fun than Microsoft's 'digital twin', but both versions miss the glory of the point.

The point is the possibility of different selves that inhabit different space-time. Why must the metaverse be in the present? Time there doesn't exist. Does it? I mean, it will, functionally, so that we all know what time the concert starts, but it won't exist in the way humans presently experience it.

Would time in a metaverse need to be a 24-hour seasonal system like Planet Earth? A metaverse of Mars would be quite a different slice of space-time.

Everyone loves the idea of time travel. In the metaverse (remember, there will be LOTS of them), and with your avatar(s), time travel becomes real in a way that could be both educational – what a great way to learn history – and inspirational – a sense of humanity as a rope slung across time.

In theory, the metaverse can be a fully imagined, unlimited, creative space.

In practice, it could turn its users into digital drug addicts.

In the outside world, maybe you live in horrible accommodation, have little money, and no power. In the metaverse, forget all that, and live your dreams. It's an extreme version of Clark Kent and Superman.

This opium of the people could be used against us, to keep us passive and accepting. To keep us distracted by phantoms and lured by online riches, like fairy gold. There's a lot of talk about crypto as the currency of the metaverse – but less talk about how we feed, clothe, and house the body wearing the magic glasses. Maybe we will be given Meta-tokens along with food stamps or a universal basic income. The super-rich might prefer this. No need for social justice when we can experience immersive social media.

There is in China now, a growing group of people who identify themselves as two-dimensionals; that is, their significant life, at work and at leisure, happens online. When that experience is not 2D but 3D, will there be any incentive to be in what we call the 'real' world?

If you can't have the child you want – create an avatar-baby, a human Tamagotchi. If you don't want a messy human relationship, build a home with an avatar-lover. Would you like a huge library, with a roaring fire, and books you can take down from the shelf to read? Would you like to get on your treadmill at home and walk the coast road in Xanadu? It will feel real – though whether that real is the strange drifting half-life of the vampire kiss, that drains away the daytime and leaves us longing for night, I do not know.

When the heavy clunky goggles improve, they will become sexy smart glasses, light to wear and unintrusive. Glasses will likely cede to brain-computer implants (BCIs). You will be your own device, and able to switch between worlds simply by thinking. But you may also be bombarded by advertising you have to deliberately 'think away' from to switch it off.

OK, it sounds crazy, but in the current world, Search Engine Optimisation is big business – how to get the keywords or images to lead to YOUR site or product, and not someone else's. Imagine an influencer you follow pushing a product. You could 'experience' the ad or hype automatically – without any clicks, because your mind has already done the clicking by tuning into some place in the metaverse where you meet your influencer – and that will be popular because it will feel like meeting someone who interests you.

If, and when, we can enter metaverses via our thoughts, there will start to be something like a dictionary of the brain – where emotions and sensations can be transmitted without language. I will be able to sense how you feel. We do that anyway, but this will be a direct experience of the emotion of another. And of course, the marketers will be able to 'tell' if you are interested in their product or service. It's a nowhere-to-hide technology coming our way.

Thomas Oxley, CEO of Synchron – a company based in

Brooklyn that is developing BCIs – discussed this 'dictionary of the brain' idea with me at the TED 2022 conference in Vancouver.

Oxley is a medical doctor and neurologist. The company's BCI implant technology has just been given FDA clearance to go ahead, as of July 2022, and the first patient, whose thoughts can turn into text, has successfully had the implant. Oxley believes that this type of surgery will become as simple and normal as LASIK eye surgery. This opens the way for healthy people to become their own device, and to connect to the internet via their brain – and it's this capacity that would allow us, potentially, to 'read' one another's emotional states, bypassing the need for language as a 'device'.

What Thomas Oxley and I argued about is whether language itself is a player – not just a communication tool. When we find the words, are we, in fact, *shaping* the thought, and therefore the feeling, rather than just communicating it? I believe so.

Language is not just a capacity – certainly not a set of instructions – it's a mental reality.

I guess, though, the hive-mind is likely to be a subset of merged reality, and non-verbal communication is inevitable. Alan Turing was fascinated by hive-mind communication back in 1950. Have a look at his paper *Computing Machinery and Intelligence*.

The metaverse will be a different kind of communication system, as well as a different kind of experienced space.

What's for sure is that the brain is not a passive organ meekly processing sense-impressions from a world 'out there'. The brain creates its own reality. The metaverse could act as a psychedelic experience that radically affects our perception of ourselves back in the 'real' world. Mind-altering drugs are being trialled for depression and mental disorder because those drugs change the brain. Anyone who has self-medicated knows the effects of cocaine or LSD or a gentle dose of cannabis. What if the chemical effect of drugs that alter our brain chemistry can be reproduced by readjusting our experience of 'reality' in other ways?

Those adapt at meditation know what it is like to leave the body behind while being centred in that same body.

Shamans in non-western cultures have always taught that the physical body remains while the astral body travels. That's what the metaverse will allow.

We know that the brain will 'believe' its meta-experiences and transmit sensations to the body – like fear or happiness or vertigo or a sense of weightlessness – but could our 3D physical body feel things in the way that we do every day?

That's being worked on right now – though the early results will be as clunky as the VR headsets. Meta is trying out a haptic glove, and has partnered with a Californian start-up company, Emerge, that has developed a device with hand-tracking sensors, using ultrasonic waves. So, when your avatar arrives at a meeting, you can squeeze the handshake, feel the high-five and the backslap, or stroke the office's digital dog.

A Japanese company backed by Sony, H2L Technologies, is developing a wristband to be used alongside goggles or glasses. Early trials allow the wearer to feel the weight of what they pick up in a virtual space, or to catch a ball. They can also feel real pain if the office's digital dog bites them.

CEO Emi Tamaki believes her company will have solved the problem of feeling-immersion in virtual space by 2029. As a small, light, woman, she isn't interested in the oversized boys' own solutions presently on offer. As we know from mobile phones and countless other objects – including car seats – men assume their body size is the template. Not so, dudes! Women in the metaverse won't want to be weighed down with what feels like diving gear. More women working on the hardware, as well as the software, will bring real differences to the user experience.

Let's say a touchy-feely metaverse is possible. This could be a huge plus for mental health issues. The talking cure could include the body, creating warm, beautiful spaces, where the nervous, sad, anxious body can be stroked, or float, be weightless, be held by another, without fear.

And, of course, we will be able to feel the hug from our friend who is far away, when we sit together by the virtual ocean, sieving the gritty sand under our haptic-touch fingers.

This technology could be used in hospitals and care homes – in any circumstances where a virtual visit that feels real will make a difference.

In a future pandemic, a metaverse you can touch will render life less lonely.

All good? Sure! But many of the articles you will read about this, stress the dangers too – and it's right to be thinking about the dangers.

The mind is a powerful freedom tool, but is also its own jailor, and when we are not jailing ourselves, there's plenty of others who will oppress us. How this plays out in the metaverse – utopia or dystopia – must be decided early and soon.

This isn't an accidental universe – we are building it from scratch. We could do better this time than all the times before when we've said let's find new territory and start again. All those utopias that went wrong straight away. The Pilgrim Fathers who landed at Plymouth Rock prioritised a church and a gaol. And strict gender roles. The land was a fresh start; the humans were the same old story. And the story is in our minds. That's the place where change has to happen, whatever universe we are in.

We know the lessons of history. We know the evil in our hearts. Can we learn from what we are and honestly start again?

Not if it's just a giant shopping mall, we can't.

Neal Stephenson's novel *Snow Crash* coined the term metaverse in 1992. The world he creates is a dystopia, cynically run for profit and power, understood by just a few. Something like Kafka's *The Castle* (1926).

In the *Matrix* movies, there are layers of virtual reality – but there is still a purely human reality: the resistance movement that operates underground. If the metaverse(s) is built according to its present plan, the stated aim is to blend levels of reality. And ultimately, to alter what we know as reality.

Does it matter? Will it matter?

Throughout the essays in *12 Bytes*, I have drawn attention to Shoshana Zuboff's question in her book *The Age of Surveillance Capitalism*: who decides? Who decides who decides?

There is basic stuff around the internet, as it is now, that needs to be decided before we can go further – particularly privacy issues and trust issues. Who we can trust out there is impossible to know. In the metaverse we will be open to every kind of deception and scam (yes, I know it's a caring sharing community – pass the sick bag with the company logo on it).

Big Tech, Meta, especially, in its Facebook and Insta incarnations, has failed miserably to do enough to stop trolling and stalking online. Being trolled and stalked by an avatar every time you enter the metaverse is just as likely as making a new friend.

White supremacists can shoot down whoever they like – it's not real, is it? But it will still be terrorism. A reign of terror against certain people, companies, charities, is a genuine threat.

Governments can't keep up with Big Tech, and have swallowed the line that legislation strangles innovation. What we really need, before the metaverse rolls out, is a global governance body, with legal weight – an organisation on the scale of the UN, or the World Bank, or NATO – open to everyone who wants a fair, clear, rules-based alternative universe.

If we do it nation by nation – or even blocs of nations – it won't be enough. I envisage a public–private partnership for global good. Why would that be so hard to achieve?

Well, OK, we know why it would be hard to achieve, but hell, are we saying we can build an alternative world, but we don't know how to co-operate with each other?

A global governance organisation could urgently address the access issue too.

According to a 2019 UNESCO report, only 55% of households have an internet connection. In the developed world, 87% of us are connected, but that drops to 19% in the world's poorest nations. The UN calculates that 37% of the world's population have never been online.[1]

If there is no money to be made for the providers, then the poor will go on being shut out of even basic internet services. That could be addressed now, with global not-for-profit governance.

Will the metaverse worsen the world's present unequal divide; who is connected and who is not? Or shall we urgently and deliberately work to connect everyone, at the same time as we are building Web 3.0?

And what about kids in the metaverse? We have only just got general global co-operation over the simple and uncontroversial fact that kids should be protected in the digital world as they are in the real world. In the metaverse, children and young people will be vulnerable in new and terrible ways – we must make sure it doesn't happen.

These are important and necessary questions – and while we are talking about access, and about kids, we should be talking about women too. Namely, where are they?

Women are not doing much of the deciding/designing of the metaworlds to come. If women don't design those worlds, then women are tourists, at best.

Feminist Met-opia anyone?

I am worried too about the increasing craziness of right-wing America, where angry voices are calling out-loud for a theocracy. A return to God that will magically end mass shootings, racial violence, and divorce. I can only speculate that too many folks who hate the tolerance and inclusivity of the best of the modern world (and that's what America used to be) watched *The Handmaid's Tale*, and thought: 'Gilead, there's the place to live.'

The Great American Desire to control the bodies of women is only the beginning of a western-style Taliban. No one of my generation expected to see a handful of religious judges on the

1 *Guardian*, 30 November 2021.

Supreme Court overturn a woman's right to choose. It has happened. There's more to come.

There is nothing to stop right-wing Godists building their own metaverse.

Scenario: your curious avatar goes to visit a virtual megachurch in the metaverse's Paradise City – built by the godly for the godly. Or even a bake-sale on behalf of refugees. Christians do charity, don't they? Cookies will then track you back into your other worlds – information-gathering and seeking to influence your behaviour.

Scenario: your avatar is found guilty of undermining the moral values of Paradise City. She is captured and tortured. She comes to you via your smart glasses, or implant, begging for help. It's not you, and you know that, but you don't feel that way – it would be the most frightening mind game.

As our online selves gain a real presence, we can do real harm – and experience real hurt, in new and ingenious ways.

And even if everything is a rose-bed, your avatar will be a permanent tracking code back to 'you' for as long as there is such a thing as 'you'. You are your own double-agent. You are the spy who spies on yourself. Don't forget that.

At the same time, there is such capacity for beauty here. For freedom and enrichment – but what's new? Humans have endlessly invented machinery and technology that could benefit the many, not the few, and always, the benefits go to the few, and the burdens, physical, mental, environmental, to the many.

Whatever we invent, our solutions soon become our problems.

There is no hi-tech solution to human stupidity and greed.

Unless, that is, we do consciously begin to seek a transhuman, and eventually a post-human future. An evolutionary choice that is in our hands. Might that mean creating a new god? A superpower intelligence? Maybe.

Maybe the freedom from physical limits that we will

experience in the metaverse will give us a sense of what a fully evolved Homo sapiens could be.

It is important that Mark Zuckerberg, and what some might view as his surveillance company now known as Meta, doesn't control the metaverse. Other big players will want their share, of course, and at present, metaverses will be like medieval walled cities – you won't be able to move your 'stuff' from one place to another. There's no virtual U-Haul yet.

Adam Mosseri, CEO of Instagram, spoke in his TED talk in 2022 about a passport system run by Meta that would allow passage of goods and services and stuff, and 'you', to happen between worlds (platforms). Sounds great, but in effect, is Meta planning to control immigration in cyberspace? There will have to be a revenue stream – it's not going to be a Good Deed.

But perhaps we will have to have identity cards in the metaverse. I don't think humans are ready for unregulated space. I hope we are ready for invented space.

As someone who makes a living out of inventing worlds, and who knows that reading a book has been the key to the lock for so many people across time and place, I believe in inventing worlds. Invent as much as you can – especially yourself; don't be a prisoner of your own facts – be the pioneer of your own fiction.

We are more than the data-sets we were trained on. There is escape.

That's not the same thing as escapism – which might be all that Meta has to offer, as the planet heats, and wars threaten what's left.

The metaverse has the potential for all of us to explore our inner space; the creative self not confined by physical or practical limits – certainly not by gender or skin colour.

Every child is a world builder. If she has no toys, she will sit on the kitchen floor, and make a world out of pots and pans. Children are born into a blended reality. Their dog and their teddy bear sleep on the same bed. The story of the wicked witch is as real as the man on the bus who pulls funny faces. We teach children what is real, and what is not, with the bizarre exception of religious teaching, where kids are encouraged to believe in a whole panoply of immortal, invisible beings, keenly interested in their welfare – at least until the sins pile up. Humans come pre-loaded for the metaverse, and its seductions of augmented, blended, virtual, and eventually, indistinguishable worlds.

And what about our human interaction with non-human systems? In the metaverse the Turing Test – can you tell it's a machine and not a person – may not be relevant.

In June 2022, Google suspended software engineer Blake Lemoine, part of the artificial intelligence development team, for going public with his claim that a Google LaMDA chatbot he was working on was sentient.

Language Model for Dialogue Applications AI is trained on vast amounts of text, and works by predicting what letters, and therefore what sense, will follow – and so it appears to respond as humans do (or better than a lot of humans do) when in fact it is stringing together plausible responses to prompts – in effect, going through all the variables at lightning speed.

Lemoine's claims have been roundly trashed. In July 2022, Google fired him for breaching confidentiality policies. Lemoine refuses to back down.

But are we missing the point here? Are we asking the wrong question? If there's a chatbot out there that humans will form a relationship with (see the essay here, 'My Bear Can Talk'), then that is more interesting, to me, than deciding whether the program is sentient.

Of course, our brains anthropomorphise any interaction. We are always looking for connection. In the metaverse, being biological isn't where it's at. Why wouldn't chatbots be a full and fascinating part of our metaverse reality? Those chatbots can be embodied – and shift their embodiment, just as we can.

As I've said elsewhere in these essays, a significant number of people have always, and do still, experience their most profound relationship with a non-bio entity that likely doesn't exist: a sky-god.

The metaverse will be a place of challenge. A place where assumptions won't help us. It's a place to be open-minded and see what human feels like when it doesn't feel confined to a single physical self in a time-bubble.

Perhaps the metaverse is the next step on our evolutionary journey – but towards what? Not one thing, not one time, not one place, not one experience, not one life. Not one self. More life into a time without boundaries?

And there is one last thing for me to say – at least for now – given how fast this world is moving.

A metaverse is a persistent, alternate world. You go there as an avatar. This seems to be what Shakespeare understood in those plays of his known as 'comedies' (because they have a happy ending, kind of). Think about *A Midsummer Night's Dream*, or *As You Like It*, where characters who are, and aren't, themselves (avatars twice over, because they are actors playing characters who aren't themselves) find themselves in a forest – a look-alike world where the rules are slightly different. We could talk about *All's Well That Ends Well*, a shipwreck to shore situation, where characters in disguise mix with the 'local' inhabitants, just as human avatars will mix with avatars specially created for the metaverse.

Above all, there is *The Tempest*. On an island (metaverse), Prospero lures in the humans who must discover that they are really playing a part – and always were. Shakespeare sums it up pretty well:

> Our revels now are ended. These our actors,
> As I foretold you, were all spirits and
> Are melted into air, into thin air:
> And, like the baseless fabric of this vision,

The cloud-capp'd towers, the gorgeous palaces,
The solemn temples, the great globe itself,
Yea, all which it inherit, shall dissolve
And, like this insubstantial pageant faded,
Leave not a rack behind. We are such stuff
As dreams are made on, and our little life
Is rounded with a sleep.

Do you see? Not solid. Not 3D. Not biological. The only thing Shakespeare got wrong is that our revels are ended. I expect they are about to begin.

Selected Bibliography

These essays are not a research project; they are an exploration. My travel kit is what I am – including the contents of my head – and that would be a long list of books by now. Getting older has its benefits.

Essay by essay there are many books cited, but here are some of the sources I found most helpful.

Love(lace) Actually

Frankenstein, Mary Shelley, 1818

Frankissstein: A Love Story, Jeanette Winterson, 2019

The Thrilling Adventures of Lovelace and Babbage: The (Mostly) True Story of the First Computer, Sydney Padua, 2015 (NOTE: FUCKING GREAT!)

Sketch of the Analytical Engine invented by Charles Babbage, Esq ... with notes by the translator, L. F. Menabrea, 1842. Extracted from the 'Scientific Memoirs' [The translator's notes signed: A.L.L. i.e. Augusta Ada King, Countess Lovelace.]

Byron: Life and Legend, Fiona MacCarthy, 2002

Romantic Outlaws: The Extraordinary Lives of Mary Wollstonecraft and her Daughter Mary Shelley, Charlotte Gordon, 2015

A Vindication of the Rights of Woman: With Strictures on Political and Moral Subjects, Mary Wollstonecraft, 1792

Rights of Man, Thomas Paine, 1791

The United States Declaration of Independence, 1776

The Social Contract, Jean-Jacques Rousseau, 1762

Enquiry Concerning Political Justice and Its Influence on Morals and Happiness, William Godwin, 1793

Cold Comfort Farm, Stella Gibbons, 1932

Hidden Figures (movie), directed by Theodore Melfi, 2016

'The Women of ENIAC' (essay), *IEEE Annals of the History of Computing*, 1996 (Interviewing 10 of the women who worked with the computer during its 10-year run)

The Creativity Code: Art and Innovation in the Age of AI, Marcus du Sautoy, 2019

Howards End, E. M. Forster, 1910

A Loom with a View

A Cyborg Manifesto, 1985, and *Staying with the Trouble*, 2016, Donna J. Haraway

The Singularity Is Near: When Humans Transcend Biology, Ray Kurzweil, 2005

The Condition of the Working Class in England, Friedrich Engels, 1845

The Communist Manifesto, Karl Marx and Friedrich Engels, 1848

The Subjection of Women, John Stuart Mill, 1869

The Making of the English Working Class, E. P. Thompson, 1963

Industry and Empire: From 1750 to the Present Day, Eric Hobsbawm, 1968

Why the West Rules – For Now, Ian Morris, 2010

Debt: The First 5000 Years, David Graeber, 2011

'The Masque of Anarchy' (poem), Percy Bysshe Shelley, 1832: 'Ye are many—they are few'

'A Short History of Enclosure in Britain' (essay), Simon Fairlie, 2009

PostCapitalism: A Guide to Our Future, Paul Mason, 2015

Capital in the Twenty-First Century, Thomas Piketty, 2013

Move Fast and Break Things: How Facebook, Google, and Amazon have cornered culture and undermined democracy, Jonathan Taplin, 2017

The Mill on the Floss, George Eliot, 1860

From Sci-fi to Wi-fi to My-Wi

Rocannon's World, Ursula K. Le Guin, 1966

The Midwich Cuckoos, John Wyndham, 1957

Brave New World, Aldous Huxley, 1932

Weaving the Web: The Original Design and Ultimate Destiny of the World Wide Web, Tim Berners-Lee, 1999

'We Can Remember It for You Wholesale' (short story), Philip K. Dick, 1966

The Age of Surveillance Capitalism: The Fight for a Human Future at the New Frontier of Power, Shoshana Zuboff, 2018.

The Four: The Hidden DNA of Amazon, Apple, Facebook, and Google, Scott Galloway, 2017

Becoming Steve Jobs: The Evolution of a Reckless Upstart, Brent Schlender and Rick Tetzeli, 2015

How Google Works, Eric Schmidt and Jonathan Rosenberg, 2014

The Art of Electronics, Paul Horowitz and Winfield Hill, 1980 (NOTE: I only bought this because I thought it said Winifred Hill – and girls don't do circuits, do they? Anyway, it's v good.)

Gnostic Know-How

The Society of Mind, Marvin Minsky, 1986

2001: A Space Odyssey, Arthur C. Clarke, 1968

Probability and the Weighing of Evidence, I. J. Good, 1950

Our Final Invention: Artificial Intelligence and the End of the Human Era, James Barrat, 2013

'Computing Machinery and Intelligence' (article), Alan Turing, 1950

The Gnostic Gospels, Elaine Pagels, 1979

The Nag Hammadi Scriptures, edited by Marvin W. Meyer, 2007

Mysterium Coniunctionis, Carl Jung, 1955

On the Origin of Species, Charles Darwin, 1859

The Odyssey, Homer

He Ain't Heavy, He's My Buddha

An Introduction to Cybernetics, W. Ross Ashby, 1956

Buddhism for Beginners, Thubten Chodron, 2001

A Simple Path: Basic Buddhist Teachings, His Holiness the Dalai Lama, 2000

The Tao of Physics: An Exploration of the Parallels Between Modern Physics and Eastern Mysticism, Fritjof Capra, 1975

The Systems View of Life: A Unifying Vision, Fritjof Capra and Pier Luigi Luisi, 2014

Wholeness and the Implicate Order, David Bohm, 1980

Reality Is Not What It Seems, Carlo Rovelli, 2014

A History of Western Philosophy, Bertrand Russell, 1945

The Sovereignty of Good, Iris Murdoch, 1970

A Little History of Philosophy, Nigel Warburton, 2011

The Symposium, Plato

On the Soul and *Poetics*, Aristotle

Aristotle's Way: How Ancient Wisdom Can Change Your Life, Edith Hall, 2018

Shakespeare's sonnets

Opticks, Isaac Newton, 1704

The Future of the Mind: The Scientific Quest to Understand, Enhance and Empower the Mind, Michio Kaku, 2014 (I am just about to read Kaku's *The God Equation: The Quest for a Theory of Everything*, 2021. I recommend all his books.)

The Age of Spiritual Machines: When Computers Exceed Human Intelligence, Ray Kurzweil, 1999

Novacene: The Coming Age of Hyperintelligence, James Lovelock, 2019 (and everything he has written)

Coal-Fired Vampire

Epic of Gilgamesh (world's earliest surviving text)

Dracula, Bram Stoker, 1897

Interview with the Vampire, Anne Rice, 1976

The Twilight saga, Stephenie Meyer 2005–20

The Picture of Dorian Gray, Oscar Wilde, 1890

Faust, Goethe, 1808

The Divine Comedy, Dante, 1472

'Piers Plowman' (poem), William Langland, 1370–90

'The Vampyre' (short story), John William Polidori, 1819

How to Create a Mind: The Secret of Human Thought Revealed, Ray Kurzweil, 2012

'Transhumanism' (article), Julian Huxley, 1968

Superintelligence: Paths, Dangers, Strategies, Nick Bostrom, 2014

To Be a Machine: Adventures Among Cyborgs, Utopians, Hackers, and the Futurists Solving the Modest Problem of Death, Mark O'Connell, 2017

Selected poems of Andrew Marvell, 1995 (I should have included 'To His Coy Mistress' in the essay. All about death – and what to do about it.)

Also have a look at the painting 'The Harrowing of Hell' by an imitator of Hieronymus Bosch – if you Google it, it comes up as Shop Harrowing of Hell, so I guess even redemption is a consumer experience now.

Hot for a Bot

It's all in the essay, folks! No need to repeat it here. Just one thing – Marge Piercy's novel *He, She and It*, a cyborg love affair, where the man is the cyborg, and the woman is in charge. That was 1991 – the technology is moving forward but our minds are not.

My Bear Can Talk

All the Winnie the Pooh books! A. A. Milne, 1926

Goodnight Moon, Margaret Wise Brown, 1947

The Child and the Family: First Relationships, 1957, *The Child, the Family, and the Outside World*, 1964, and *Playing and Reality*, 1971, Donald Winnicott

I, Robot, Isaac Asimov, 1950

I Sing the Body Electric!, Ray Bradbury, 1969

Do Androids Dream of Electric Sheep?, Philip K Dick, 1968

R.U.R.: Rossum's Universal Robots, Karel Čapek, 1920

AI: Its Nature and Future, Margaret A. Boden, 2016

My Robot Gets Me: How Social Design Can Make New Products More Human, Carla Diana, 2021

Fuck the Binary

The Descent of Man, and Selection in Relation to Sex, Charles Darwin, 1871

Hereditary Genius, Francis Galton, 1869

An Essay Concerning Human Understanding, John Locke, 1689

Orlando: A Biography, Virginia Woolf, 1928

The Left Hand of Darkness, Ursula K. Le Guin, 1969

The Handmaid's Tale, Margaret Atwood, 1985

Written on the Body, 1992, and *The Powerbook*, 2000, Jeanette Winterson

Freshwater, Akwaeke Emezi, 2018

Gender Trouble: Feminism and the Subversion of Identity, Judith Butler, 1990

The Hélène Cixous Reader, Ed. Susan Sellers, 1994

The Dialectic of Sex: The Case for Feminist Revolution, Shulamith Firestone, 1970

Sapiens: A Brief History of Humankind, Yuval Noah Harari, 2011

Invisible Women: Exposing Data Bias in a World Designed for Men, Caroline Criado Perez, 2019

The *I-Ching*

Testosterone Rex: Myths of Sex, Science, and Society, Cordelia Fine, 2017 (and everything she has written and will write)

The Gendered Brain: The New Neuroscience That Shatters the Myth of the Female Brain, Gina Rippon, 2019

The Future Isn't Female

Unlocking the Clubhouse: Women in Computing, Jane Margolis and Allan Fisher, 2002

Programmed Inequality: How Britain Discarded Women Technologists and Lost Its Edge in Computing, Marie Hicks, 2017

Algorithims of Oppression: How Search Engines Reinforce Racism, Safiya Umoja Noble, 2018

The Glass Universe: How the Ladies of the Harvard Observatory Took the Measure of the Stars, Dava Sobel, 2016

Let it Go: My Extraordinary Story – from Refugee to Entrepreneur to Philanthropist, the memoir of Dame Stephanie Shirley, 2012 (If you don't have time for this, just find her TED Talk.)

Uncanny Valley, Anna Wiener, 2020

The Second Sex, Simone de Beauvoir, 1949

Hackers: Heroes of the Computer Revolution, Steven Levy, 1984

Psychology of Crowds, Gustave Le Bon, 1896

Lean In: Women, Work, and the Will to Lead, Sheryl Sandberg, 2013

Difficult Women: A History of Feminism in 11 Fights, Helen Lewis, 2020

A Room of One's Own, Virginia Woolf, 1929

Your Computer Is on Fire, various editors, 2021 (haven't read this at time of going to press but looks great)

The Blank Slate: The Modern Denial of Human Nature, Steven Pinker, 2002

Of Woman Born: Motherhood as Experience and Institution, Adrienne Rich, 1976

The Better Half: On the Genetic Superiority of Women, Sharon Moalem, 2020

Jurassic Car Park

Nineteen Eighty-Four, George Orwell, 1949

The War of the Worlds, H. G. Wells, 1898

People, Power, and Profits: Progressive Capitalism for an Age of Discontent, Joseph Stiglitz, 2019

The Sixth Extinction: An Unnatural History, Elizabeth Kolbert, 2014

Utopia for Realists: The Case for a Universal Basic Income, Open Borders, and a 15-hour Workweek, 2014, and *Humankind: A Hopeful History*, 2019, Rutger Bregman

Notes from an Apocalypse: A Personal Journey to the End of the World and Back, Mark O'Connell, 2020

The Better Angels of Our Nature: Why Violence Has Declined, Steven Pinker, 2011

Blockchain Chicken Farm: And Other Stories of Tech in China's Countryside, Xiaowei Wang, 2020

Life 3.0: Being Human in the Age of Artificial Intelligence, Max Tegmark, 2017

The Alignment Problem: How Can Machines Learn Human Values?, Brian Christian, 2021

I Love, Therefore I Am

There is no reading list. There is everything you are.

Illustration and Text Credits

But the truth is I had no plan. All I knew was that the weekend seemed like a safe time to talk. My wife and I would wake up late, for once, and lie together as long as we could, before the demands of daughter and dog forced us to get out of bed. We always treasured those rare moments that we had to ourselves: the toes of her left foot tucked between my ankles, the warm weight of my hand on hers, so familiar that it was almost imperceptible, a breeze through the half-open window, sun mottling the wall behind our heads. I wouldn't have to jam my explanations into the margins of routine; there would be no daily stresses to infringe upon what needed to be said.

It was supposed to be a gorgeous day, finally—the first really warm day of the long-awaited spring. My wife and I would sit together on our front porch and talk about the future, figure things out. I could help her understand. I could soothe her, make my words right, promise her that everything would turn out fine for us in the end.

I just hadn't counted on the school project.

■ ■ ■

Let's see us again, that day.

I remember the three of us working together in the dining room.

I remember sun streaming through the venetian blinds onto the table, slices of light cutting over ill-formed papier-mâché mountains.

I remember the vase of orange tulips, four days in water, opening their petals to reveal the black dust within.

I remember Beau poking around at our ankles, looking for edible morsels as always. He must have eaten something bad the night before, because early that morning he had vomited a thin yellow gruel onto my daughter's bed. Now her duvet cover was in the wash, the rhythmic bump of the dryer balls like the beat of a frightened heart as the laundry tumbled over and over.

I remember my wife, sun on her dark hair as she bent over her work. I remember pausing behind her to knead her shoulders, my palms absorbing the warmth of her skin under her sheer cotton shirt. I remember her moving around the room humming, watercolor brush like a conductor's baton in her hand, a lightness and looseness in her gestures that I had not seen in a long time.

I remember a repairman was supposed to stop by that afternoon to take a look at the dishwasher, which had been leaking since Monday. The lawn needed mowing. The neighbor's Callery pear trees had bloomed, saturating the block with the noxious smell of semen. There was a leak behind the fireplace that had developed sometime in February during a late-season snowstorm, and the area over the mantel had puckered, tea-colored stains spreading outward from the seam below the ceiling. White dust from the rotted drywall had seeped out, blooming over the bricks of the fireplace. That entire wall would have to be torn down and rebuilt.

I remember Tannhäuser throwing himself at Elisabeth's feet and refusing to tell her where he's been, no matter how much she begs. And I remember the way the joyful song of their reunion was lost to the roar of static.

Late in the afternoon, I tried to help my daughter with the unwritten essay.

Okay, I said, opening a new document on my laptop. Let's get this done so you can go see your aunts and your grandmother. They're all expecting you.

My girl slumped in the chair beside me.

This shouldn't be too hard to write, I said. You just have to write about what you think the call of the wild is.

I don't know what it is, my daughter said, pouting. I don't know what we're supposed to say.

There's nothing that you're *supposed* to say. There's no one right answer. It can be whatever you think it is.

Well, I don't know.

Come on, you can do this. Just think about it.

My daughter stared into space, silent. I rapped the table with my palm.

What?

Sit up.

Fine.

Could it be nature? Or instinct?

Maybe. I don't know.

You read the book, didn't you?

Yes.

Okay. Let's try the other essay question instead. In what way is the novel fervently American?

I can't answer that one either.

Why not?

Because I don't know what that means!

You just need to come up with a thesis, I said. You know how to do that, right?

No.

No?

They never taught us how.

That's impossible.

They always say to just write down the theme. And connect it to the main idea.

What does that even mean?

I don't know.

If you don't understand something they're teaching you, you need to ask more questions.

Okay.

Look, your thesis is essentially an argument. So what's your argument?

My daughter sighed loudly, dramatically. This is the most boring book I've ever read, she said. That's my argument.

That's not helpful.

She looked away and let her eyes settle on the chaos of the table and the unfinished game. Who had taught her to wait until the last minute like this? Certainly I hadn't.

Sweetheart, I began. How long have you known about this project?

No response.

Weeks? I pressed. Months?

She wouldn't speak. My Princess Turandot. She was freezing me out now, as if she couldn't hear me, or didn't care to. As if I weren't there at all.

Look, it's not so hard, I said. We decide what the call of the wild is, and then we say why we think that. So what do you think it could be?

I don't know. Nature or something. What was the other thing you said before?

Instinct.

Or instinct, I guess. Whatever you think it is.

Come on, you can do better than that. I know you're capable of doing brilliant things.

But Daddy, I don't *know*!

She was breathing hard now, whimpering between each exhaled gasp, her hands clenched into fists. I watched her for a while, waiting for her to collect herself, but she didn't.

Calm down. Just breathe.

I can't!

All right, all right, I said. Just relax. I'll help you write it.

She went slack then, her face as still and pale as a doll's.

Trust me, I told her. It's going to be great. *You're* going to be great.

My wife had left the room at some point—to do what, I didn't know. She seemed relaxed. I was surprised to see her so apparently unbothered by the looming deadline we were facing. Perhaps she was just trying to keep us calm. She'd always been the one who maintained order and smoothed the course, no matter what problems might befall us.

Beau, on the other hand, seemed jumpier than usual. He kept bristling at hints of sound. He stared at the squirrels that stood still on the lawn, growled low when they darted away.

We were still a real family that day, doing the exact kinds of things that real families do. The project would get done—the thesis would be generated, the essay would be written, the verdant paint on the mountains would dry, the clay wolves would find their homes among them—and we would all be relieved. My daughter would have her movie night in the haven of my mother's house, and my wife and I would have a Saturday evening to ourselves. It would all be fine.

All I had ever wanted was to do everything the right way. All I had wanted was the love and stability that had never been mine all through my miserable childhood. I had wanted to build a shell around my girls, so that they would never have to suffer as I had.

I had once truly believed that I could have a better life—I thought I had seen only the bright colors of happiness arrayed before me like a great garden of my own making. But I had been wrong. The good times had already come and gone, and the garden had given up its blooms before I knew to harvest them. I had destroyed our lives, just like my father had destroyed the lives of his own wife and children. This house I'd built with my own hands was decaying now, falling to pieces, and no amount of light or air could save it.

Unless.

Wasn't I still the hero of this story? Who was I if I couldn't shepherd us safely through a crisis? I had to make things right, for their sake. They were everything to me—they would always be everything.

I knew that we desperately needed an evening alone, my wife and I. As soon as we finished enough work on the project we could get our daughter out of there, pack her off to her grandmother and aunts. Then my wife and I would sit down together and I would finally tell her what had happened. I would open a bottle of wine and wait until she'd taken her first smiling, ruddy-tongued sips before saying anything.

Then I would lean toward her, brush the hair out of her face with the gentle blade of my hand, kiss the soft corner of her mouth, encase her in my arms so that she couldn't move, so that I could feel the sympathetic flutter of her sighs.

It's all a misunderstanding, I would say, her head warm and silent against my chest. You know me, my love. You know I would never do those things. Could never.

With time I could make her see that our situation was just like that of poor Tannhäuser. He lost control of himself and sang to the assembled crowd a forbidden song of profane, ecstatic love. He was condemned to die, and certainly would have if not for Elisabeth. She pleads for mercy on his behalf, promising that the sinner she loves will achieve salvation through atonement. And she vows that if he fails, she will plead his case to God herself.

Her love for that poor sinner, that poor bastard—that was the kind of love my wife had for me. More powerful than any disaster. Insoluble. Indelible. I would beg for forgiveness, and my wife would forgive me. Because she loved me.

I held on to those thoughts, nursing them like a puncture wound in my chest as my girls and I continued our work, hymns of forgiveness assaulting us from the stereo.

■ ■ ■

The dishwasher repairman never came. It was after five now, and we were moving into the next stage of the evening. The dryer had completed its cycle and the duvet cover was a warm and pristine cloud of white once more. Beau snuffed and paced around the back door as he waited for his dinner. The little clay wolves had been glued into place among the snowcapped mountains of the board game, and we had composed a sufficiently robust outline unpacking the conflict between instinct and civilization in the novel, complete with key quotes for evidence. The essay would write itself in the morning.

The opera had ended, the pilgrims overwhelming us with their last breathtaking chorus, our speakers crackling with the final ferocious tremolo of the violins as I wiped my eyes with the back of my hand. I always cried when I listened to that final scene. My wife stretched out on the sofa and watched my daughter and me as we made our preparations to leave.

Don't forget your coat, darling.

It's so warm out, Mommy. I don't need one.

It will get cold later. You'll be cold.

No I won't.

My wife sighed, but she was smiling.

Come here then.

My daughter trotted over to the sofa, then bent and kissed my wife on her forehead with theatrical care and affection, as if she were the mother putting her little child to bed.

Au revoir, mon chéri.

À bientôt.

I didn't know it then, of course—none of us knew—but it was the last time my girls would ever see each other.

We were out of the house by six. The drive was uneventful, just the usual traverse down the sunbaked spine of the island— traffic slowing near the exit for the mall, then speeding up again as we approached the South Shore. The breeze became cooler, more forceful, and full of sea, gusts beating their way into the car through our half-open windows. My daughter crossed her thin goose-fleshed arms, but still wouldn't admit that she was cold, shaking her head when I asked.

I had made this journey south so many times, more times than I could possibly count. But though it was all so familiar, today I found myself noticing and appreciating each detail of the drive, especially as we drew closer to our destination: the old town with its diners and car dealerships, each landmark as decrepit and enduring as a Roman ruin; the winding lane flanked by high hedges, the shadowy passage under the canopy of trees that hid our crumbling Victorian dollhouse from the street.

Perhaps I knew that these details would matter later, that it was important to remember them.

My sisters were waiting for us on the front porch. They were bare-legged in garish caftans, sipping from crystal goblets filled with crushed ice, cut fruit, and wine. They waved, leaning forward in their wicker rocking chairs as we exited the car.

Look who's here! Kit hooted.

My daughter raced up the wooden steps, but I hung back, leaning against the porch column.

We're so glad you could make it, Miss Call of the Wild.

Oh my *god*, my daughter moaned, collapsing into the empty rocker between my sisters. I thought we'd *never* finish it!

Is it all done?

No, I said. She's got more to do tomorrow.

That's why I can't sleep over, my daughter said, pouting. It's not fair.

Well, that's all right, Deedee said, waving everyone's disappointment away. You certainly don't have to think about it tonight.

What are we going to watch? What are we going to eat?

Whatever you want, baby girl.

Can we get sushi? Please?

Not Domino's?

Gross.

Oh, you used to love the cheesy bread, Kit teased. You used to dip it in the blue cheese sauce that came with the buffalo wings.

Not since, like, forever. That stuff is so bad for you.

One of these days you're going to get so skinny you'll disappear.

I listened to them banter. Their chatter was so light, so unencumbered by any kind of care. They would pass an evening eating salted edamame and spicy tuna rolls, watching the sort of movies that my daughter wasn't allowed to watch at home—movies about murderous cults of teenaged witches, or Shakespearean adaptations set in boarding schools. Maybe my sisters would read her cards, see what they could divine about her future. It would be a lovely evening, and when she finally came home that night, everything would be different.

How's Ma? I asked suddenly.

The girls turned to look at me, surprised to see me still standing there.

She's good, said Deedee. It's been a good day.

Do you want to see her? Kit asked. She's up in her room, but I'm pretty sure she's awake.

I saw her then, there in the front hall. She's still in her pajamas, her messy hair leavened with morning light. Some crooner is crackling on the record player. Our father has his bearish arms locked around our mother's waist, and he is dipping her, burping nonsense into her neck as she throws her head back, away from him, casting her wild eyes around the room in search of aid, in search of us. And we are watching from the stairs, the twins shrieking with helpless, thoughtless laughter as Evie takes my hand and squeezes it until our knuckles crack.

I shook my head, shook the scene out of my eyes.

No, I said quickly. That's all right. Just give her my love.

But I still lingered, listening as the engine of their cheerful conversation turned over and restarted itself, continuing on down the road without me. I stood there waiting to say good-bye until I was sure I wouldn't choke on my words.

I'm heading out, I said finally, turning to go. Call me when you're ready to come home.

Okay, Daddy.

Not too late, sweetheart.

I know.

On my way back I made a few stops. First, the good wine store in our charming little downtown, where I took my time choosing two bottles from their extensive French selection—a thirty-three-dollar organic rosé from Provence, and a very fine sixty-four-dollar Bordeaux that we'd enjoyed a year or two ago on someone's birthday. My birthday or my wife's, I couldn't recall. Or maybe it had been Christmas. I wanted to remember, but I couldn't.

Next, I stopped in at the bakery down the street, where I bought their last baguette and a raspberry lemon tart. It was the kind of dessert that my wife liked, more sour than sweet. I got a full-size tart, so that there would be enough left over for my daughter when she came home later that night. At the last minute I asked for three croissants. For the morning.

Then, using my debit card, I swiped myself into the vestibule of my bank, where I took out four hundred dollars in twenties from the ATM. There was nothing unusual about this. I always kept cash in the house—some in my sock drawer, some under the mattress—and after my wanderings this week I needed to replenish my supply.

I made it to the florist just before it closed. The dark elfin girl behind the counter helped me make a beautiful bouquet of hydrangeas—pearl white and palest blue, like something a bride would carry.

Just make sure you put them in lots of cold water, the girl called after me. Otherwise they'll be wilted by the morning.

I nodded, waved my thanks and good-bye without looking back. Outside it was evening—the warmth of the day had gone, and the streets of our town were gray and empty in the sudden gloaming.

There was no putting it off any longer. It was time for me to go home.

■ ■ ■

At this point I must pause to address some concerns, which are beginning to take on an increasingly uncomfortable and burdensome weight in my mind. Specifically, I am concerned that despite my best efforts to tell my story honestly, I may be missing the mark. Not telling it the way it should be told. Not coming off quite right. Owing to my present state of complete misery and remorse, I fear that I've made things look a certain way—cast things in a certain light, so to speak—and I know that way of looking is not entirely faithful to the reality of what actually happened, or how my girls and I actually lived our life together. As I've said before, I know it's impossible to fully capture the truth of it all, let alone communicate that truth to people who never even knew us. In the retelling, it's far too easy to miss the heart of it. But that doesn't stop me from wanting to get it right.

The truth is, my wife and I did make it to France. One summer, the year before our daughter was born. We spent a few days in Paris with her parents, days that I recall were a bit tense but pleasant as we got over our jet lag in their cramped and pungent apartment. Then the two of us took a train down to Aix-en-Provence, where my wife had gone to university. We rented an apartment with a terrace in a building that dated back to the Middle Ages, and we passed a romantic week wandering the narrow cobblestone lanes, hiking up to the scenic lookout point where Cézanne painted Mont Sainte-Victoire, drinking glasses of pallid, one-euro rosé in plazas while children kicked soccer balls and shouted to one another as dusk descended through the canopy of trees overhead.

I honestly don't remember much about the trip. I haven't even mentioned it until now because it hasn't seemed important, certainly not in the grand scheme of our life together. I do remember that my wife cried often while we were there. She cried when her mother served us couscous for dinner, a childhood favorite. She cried when I took a picture of her on the doorstep of the building where she'd lived as a student. She cried after we made love on our tiny terrace one night, her halting groans of pleasure giving way to sobs that were drowned out by the shouts of revelers below.

Whenever I asked her why she was crying, she would say that she was just so happy.

We meant to go back someday as a family, when our daughter was old enough to appreciate it. I know we would have eventually, maybe even as soon as the following summer. It would have been the perfect time—that elusive sweet spot at the beginning of adolescence, before our girl could finish growing up and leave us behind.

There was no reason we couldn't have.

■ ■ ■

When I got home I found my wife asleep in the living room where I'd left her, the dog nestled in the gap between her knees, his head resting on her thigh. They snored together, quietly and companionably. I stood in the doorway and watched them for a while, then lay down on the floor beside them, my spine pressed to the rug. I closed my own eyes and waited, letting the long minutes pass. I was reluctant to bring my wife back into wakefulness, because I knew what would happen when I did, and who knew when we would ever be at peace like this again?

Finally I roused myself. I knelt beside my wife, put out my hand, and stroked her hair and neck until she stirred and let out a soft little gasp before seeing me and smiling. She always made that sound when I woke her, no matter how gentle I tried to be. Looking at her now, I knew it wasn't time yet.

Let me cook for us tonight, I murmured. You're too tired.

Eyes closed once more, she nodded in assent. In another moment she was dozing again.

There was a flank steak in the freezer. I set it in the sink under lukewarm running water to defrost, and then I gathered ingredients to make a salad. That and the baguette I'd bought would have to serve us. I opened the wine to let it breathe. The bottle I'd picked out really deserved a special meal, but we would have to live with what I could make. I kept the radio off—for once I didn't feel like listening to anything. I cleared everything out of the dining room, moving the still-drying board game to the coffee table in the living room. Then I began to make dinner.

I find it difficult to describe this part of it. These hours. The meal I made for us, the dinner we had together that night. I find this period much harder to describe than many of the other things that happened. I can't tell you why.

I can tell you how to cook a steak, though—how I cooked our steak that night. Once the meat is defrosted and brought up to room temperature you blot it with paper towels, pressing the flesh until it's dry. Then you season aggressively with salt and pepper—if you think you're using too much, that's probably just the right amount. Heat a pan on high for a few minutes, until it starts to smoke—cast iron is best if you have it, which you should. Throw the steak in the pan and cook for two minutes on each side, not a second more. Set it on a cutting board and let it rest for another minute or so while you finish setting the table. Then carve. If you follow these steps it will be an impeccable medium rare.

That was the way my wife liked it.

I cut into the meat with our best carving knife, that first slice of the blade parting the charred exterior to reveal the rich ruby within. Perfect, as usual. When I finished getting everything ready I called to my wife, and we sat down at the dining room table together.

Things began to go wrong almost immediately. My torso suddenly tingled and burned with an uncontrollable itch, as if I had rolled around in poison sumac, or a colony of biting ants. I scratched at myself, pinching and clawing at my breast, then stood, nearly knocking over my chair.

Excuse me, I said, backing away from the table. I'm sorry, love.

Darling? my wife called after me. What's wrong?

Nothing! I responded, my voice light. Nothing's wrong.

But darling—

I just need a moment!

I locked myself in the downstairs bathroom and stripped off my shirt, fully expecting to look into the mirror and see a raw and vicious rash overtaking and disfiguring me—but there was nothing at all.

It was just me. Just my body.

I splashed cold water on my chest and neck, slapped my bare skin hard enough to leave a flush of handprints. Naked from the waist up, I stared at myself in the mirror for a long time, waiting until my lungs stopped pulsing grotesquely between the barriers of my ribs, waiting until I could control my face again.

Then I put my shirt back on and returned to my wife.

Is everything all right?

Yes, I said evenly. Everything is fine.

I sat down at the table again and began to cut my meat into pieces, but I found I couldn't bring myself to eat them. My tongue felt heavy and distended in my mouth; the wine tasted unbearably sour, but I continued drinking it. The itching had subsided, but a nauseating shiver kept passing through my body, a subtle convulsion that felt almost sexual. My wife was talking, but I couldn't hear or understand a word that she was saying. I was in that deaf and distant place again, watching my life from behind thick glass. Impotent.

Just begin somehow, I told myself. *Just say something. Just say her name.*

Darling?

I looked up. She was watching me closely, her doelike eyes dark with concern.

What's wrong?

Nothing, I said. I told you there's nothing.

Why aren't you eating?

I looked down at my plate. The shredded, exsanguinated flesh looked like something you would feed to a dog. I choked back a long swallow of wine. I opened my mouth, then closed it again. I was afraid I might begin to sob, or scream. With a sigh I leaned forward and took my wife's hands, kneaded her wrists with my thumbs, tried to find the words for what I had to tell her.

What's wrong, darling? she murmured. You can tell me.

I shook my head. I couldn't.

I'm sorry, love, I said. I'm sorry.

I closed my eyes and took a breath.

I'm just so tired.

We were silent for a while. My shivering intensified—I knew I had to get my body under control, force it through its motions. Command it.

Let's go upstairs, I said suddenly. Let's go to bed.

Right now?

Yes, I said, still holding her hands, pulling them to me. I've wanted you all day.

We should clean up first.

No we shouldn't, I said.

But—you're shaking.

I know. I want you right now.

There was something uncertain in her face now. A sense of stiffness, of hesitation. It disturbed me. Why now, when she had never questioned me like this before? If only she would trust me, I could fix this mess. I knew that everything would be all right once we were in bed together, once she was safe in my arms. I could convince her of everything then.

Come on, I said, looking into her eyes. Come.

For a long time we simply stared at one another. In those moments I feared that we were already lost to each other, but I also knew that if I could just stay strong and outlast her she would relent.

Come, I repeated.

Then she softened to me. I saw the moment of her decision—her submission—saw it in her wet and shining eyes, the way she let her body relax back into its natural state of gentleness. The way I always knew her to be. She leaned toward me, her forehead nearly touching mine, close enough that I could feel the breath of her words against my lips.

Let me take a shower, she said, her voice low. Then I'll meet you in the bedroom.

She went up, leaving the table untouched. In the kitchen, I listened to the shower run above me—I thought of drinking the rosé, but instead I filled a rocks glass with Laphroaig. It tasted like nothing. I drank one, then another, then more. Within a few minutes I had finished the bottle, which I left in the sink.

Finally I climbed the stairs to our room. I lay down on our bed, which seemed to pitch and roll beneath me. I held on to the edges of the mattress and closed my eyes against the unsteady tide that seemed to bear me ever closer to shipwreck and disaster. As I lay there, my thoughts turned, once more, to the final act of *Tannhäuser*. I thought of poor saintly Elisabeth, who has been waiting so long for her love to return to her. I thought of her hope when she sees the pilgrims approaching, her mounting dread as she searches each hooded face, her despair as she realizes her beloved sinner has failed in his quest for absolution and has not come back to her. She knows then that she will have to die to save him.

And then I knew with a terrible certainty that my wife would leave me.

How had I not realized it before now? All the evidence had been there, right in front of me. The messages I'd seen on her phone, her casual quickness in closing her browser windows these past few weeks—only now did I understand the significance of those minor clues. Even her sweet compliance was nothing more than a ploy. She had been making plans for a long time, and now everything had driven me to the point where I had to confess, and the awful shame of what I had done to us would be the final push she needed to leave me forever.

I suddenly saw the pale hart's antler of her left hand, the way it had looked in her lap that night in front of the fire—defiantly naked, unadorned by any evidence of my love, of our commitment. I saw it all now: her eyes downcast in disappointment, not modesty. Her long silences, weighted with regret. The way she carried herself in a crowd of strangers—showing off, seeking. The way her body would recoil at my mere approach. She hadn't worn my ring in months, and it was because she had already made up her mind.

She knew exactly who she had married—who I really was.

Desperate, I reasoned there had to be a way to make her stay. There had to be a way that I could convince her not to leave me.

Forgive me. Forgive me. I've made a terrible mistake. I don't deserve you. I don't even deserve to live. I'm sorry. Just forgive me. Stay. Please.

The door opened. There she was, her dark hair wet, her silk robe open to show her damp skin. She came toward me with sly eyes, smiling in her soft way.

O thou, my fair evening star.

No—a cat caught eating the canary.

I stood. Before she could speak, I put my hands on her throat.

And I swear on everything that I once held dear in my wretched life that she seemed to relax into my grip, that she surrendered herself to me. As if she'd known, somehow, all along.

The summer Evie was fourteen she was infatuated with the androgynously handsome lead singer of a mediocre British goth-rock band. She bought all of the band's albums on both vinyl and cassette and scrawled their brooding lyrics over every surface she could touch—her notebooks, her bedroom walls, the toes and soles of her sneakers. Even her underwear was marked by this man's words, obscure slant-rhymed couplets about drugs and failed relationships penned onto the cotton. I had never seen a lust like that before, had never known that someone could be so hungry with such impossible desire, least of all my older sister.

She had never been like that in our own games.

In August the singer's face and naked torso appeared on the cover of a major popular music magazine. The object of my sister's obsession lay in rumpled white sheets, his sharp cheekbones kissed by late-morning light, his dark hair tousled perhaps in agitation, perhaps by the hands of an unseen lover. His rakish eyes flashed, pleading with you to come back to bed, to stay a bit longer. He bit his lip, waiting. To Evie, it was a picture of pure sex.

The issue sold out immediately. After two weeks of searching every newsstand and bookstore within a bikeable radius of our house, Evie walked into our public library wearing one of our father's shirts, an old chambray oxford that was much too big for her. A few minutes later, she walked out with the library's copy of the magazine tucked into the front of her denim shorts, her beloved's face pressed against her stomach and hips, finally hers.

But Evie hated the rest of the cover. She was repulsed by the red and black text that advertised other bands she hated and stories she would not read. All the headlines and teasers were pollutants, and they soiled the perfection of her love.

So she tried to free him from his bonds.

We sat on her bed and I watched as she took a pair of our mother's nail scissors and began to snip carefully at the cover, pruning the words away, leaving the celestial face and body intact.

Trust me, she whispered to the lover cradled in her lap. This will be so much better.

Maybe I shifted then, trying to get more comfortable, or to get a better look. Or maybe Evie was the one who moved, crossing her legs beneath her or propping herself on an elbow.

Whatever the reason, in the next moment my sister's hand slipped, and before she could stop herself she had sliced a thick scar through her beloved's rosy left nipple.

For a long moment Evie was completely still and silent. I watched her pale face fearfully, waiting for her reaction. She let herself fall backward onto her pillows, lay as though she were dead—the same way she had been when I saw her with our father. It was as if she were not herself, not Evie anymore, just someone else's body that didn't move.

But then she raised herself up with a sharp gasp and began to howl.

It was a hoarse moan that tightened into a shriek and went on and on, a sound worse than anything I'd ever heard. Her face was grim and horrible, lit with an indescribable passion, and she gripped the scissors until her fingertips lost their blood and she slashed and shredded the image of the one she loved, the one she could not have and would never have, until all that remained were a few handfuls of ragged scraps spilled over the bed. All this time I watched her—even when she screamed at me to stop looking at her, screamed at me to go and leave her alone, I didn't move.

I stayed—not consoling her, not even trying to calm her—just watching her. And when my sister had finally spent her rage we listened to her breathing rend the silence around us, each labored inhale like something too heavy that she'd been forced to drag too far for too long.

She was still breathing. I could still hear it.

When I tried to stand up she grabbed at me, her hand clutching at my ankle in a violent spasm, but I kicked myself free. I pulled the sheets and comforter off the bed, threw them down into a pile on the floor, covering everything but her bare feet.

I still could have stopped, then. But really, I couldn't have.

The billy club was under my side of the bed. It had been there for weeks, growing a delicate skin of dust. I reached for it now, wiped it with the hem of my T-shirt. The collar had been stretched out beyond saving, and blood from my nose had stained the cotton. I would have to change and throw the garment away.

With the club I hit what was under the pile of bedding until it stopped moving, until the sound of the blows became muffled and wet. Then I lay down on the bed and blacked out.

When I woke up again she had come back. She stood at the foot of the bed, still wearing her robe. We looked at each other in the gray light of the early morning that hung like smoke in the room.

I thought you left, I said.

She shrugged, made a vague gesture.

You came back.

She didn't answer me right away. She kept looking at me, kept herself very still, her eyes a flat and unblinking black. She held her breath.

I had to, she said finally. I left all my things.

Your things?

She pointed to the rumpled pile of bedding on the floor.

But that's not—I said. That's—no. That isn't yours. Those aren't your things.

She shook her head then, slowly touched her fingertips to her throat.

Then what is it?

A phone was going off somewhere. A hard buzzing against wood, loud digital chimes ascending and descending. My arms and legs jerked and I was awake again. My eyes ranged over the blank ceiling, the black windows with their curtains drawn back, exposing me to the night. For a moment I didn't know where I was—it was possible I had been unconscious for days.

It was cold. There was a bad smell in the room that I couldn't name.

The phone was still going off, the sound lodged somewhere deep inside my head. No. It was somewhere close to me. I threw my hand back behind my head, fumbled blindly over the cluttered surface of the bedside table. There. Holding it to my face, I realized that it wasn't my phone—it was someone else's—but I answered it anyway, my numb fingers probing the screen in a pattern I didn't know I remembered.

Yes, I said. It's me.

I was convinced that I was still asleep. That I was way, way down in the deepest part of a dream. That I hadn't even begun to make my way back to the surface.

Daddy?

Her voice brought me back. I caught my breath.

Yes, sweetheart.

penguin.co.uk/vintage

Why are you on Mommy's phone?

My girl, I thought. *My girl.*

She's asleep, love, I said. What's up?

Oh. Well, can you pick me up now?

Now everything else was starting to come back. It was all beginning to surface, dead fish floating up from deep water. I avoided looking at the floor, even though I knew what was there.

Of course, dear, I said carefully. I'm leaving now. I'll be there very soon.

The phone went silent against my ear. I regarded the device for a few minutes, turning it over in my hand, trying to decide what to do with it. Finally I powered it down and put it in my pocket. Perhaps I would throw it away later, somewhere else.

I changed out of my ruined T-shirt. This too I hesitated over before leaving it balled up on the bathroom floor—I would deal with it later. I put on an old blue-and-gray flannel, a garment I almost never wore unless I was sick or doing work around the house. I turned out the light in the bedroom and shut the door behind me.

Downstairs I spent a while searching for my daughter's winter coat, the pink puffer one. She would need it now that it had gotten so cold out; before I had powered the phone down I had seen the weather icon on the screen indicating that the temperature had dropped down into the forties. It seemed unthinkable that only a few hours ago it had felt like high summer.

I left everything the way it was—the abandoned meal on the dining room table, the school project in the living room—only turning off the lights as I made my way through each room. It would all be dealt with later, somehow. I finally found the pink puffer coat hanging on a hook near the side door, under a dark olive trench coat and some other jackets and sweaters that had been worn by other people more recently. The dog's leash was there as well—I took it in hand, then put on my own coat.

I believe it was at this point that I realized I was carrying the billy club around with me. It hung loosely at my side, an extension of my right arm. I didn't examine it as closely as I had the phone or my shirt. I simply set it down on the kitchen counter for a moment, then called to my dog.

He was upstairs, pawing at the bedroom door, yipping loudly and fretfully. When he heard me calling he hurried to the landing to peer down beseechingly at me. I crawled up the stairs on my hands and knees, stayed crouched down low and beckoned to my dog, cooing endearments.

Come here, I murmured, reaching out for him. Good boy, good boy. Come here.

Finally he approached me with a halting gait, came close enough to let me fondle his bowed head. He was a small dog, had probably been the runt of the litter. Years ago, when we'd first brought him home from the shelter, he'd crawled under the sofa and slept there for an hour before finally allowing us to coax him out, to hold and love him.

Come on, boy, I said. Let's go out.

I notched my fingers under his collar and clipped the retractable leash on, then led him downstairs—his curly head hesitating over the precipice of each step, his paws making their soft, syncopated thuds as we descended. We went into the kitchen.

I was careless—I had given the leash too much slack and held it too loosely, and when my dog saw the club in my hand he jerked away and out of my grip, the leash retracting with a snap, its plastic handle clattering after him as he scuttled away. I chased him for several awkward, lunging laps around the kitchen island—clockwise, then counterclockwise, then clockwise again—each swing of the club a moment too late, a beat too slow. Finally I feinted right, then darted left, snatching the handle of the leash just before it got away from me again.

Breathing harder now, I stood and wound the thin cord of the leash around and around my wrist, pulling my dog to me across the hardwood floor, just close enough so that I could land the blow.

■ ■ ■

I don't remember leaving my house, or any details of the drive south. At some point I glanced behind me and saw that I had placed the billy club on the backseat, partly covering it with my daughter's pink coat. The bright beams of the streetlights kept slicing through the car, jumping over the mound of the coat and club. I faced forward again and kept my eyes on the road.

I thought idly of where my daughter and I would go and how we would rebuild our life together once I had her with me again. I decided not to worry about specifics just then. Together, we would make our plans and find a way to get over our pain.

I thought of a television spot I'd seen once for Homesteader, a real estate search engine. Our agency hadn't made the ad, but I'd greatly admired it. The commercial was the kind of work I aspired to, a poignant and compelling story that stayed with the viewer long after its images had receded from the screen. No small achievement these days.

The ad's opening scene takes place in a darkened kitchen, where a man sits at the counter, looking at his laptop. The light of the screen illuminates his somber face as he searches houses for sale, clicking through slideshows of cottages with picket fences, bungalows with big trees in their front yards. He frowns.

The man's young daughter comes in wearing pajamas, a bedraggled teddy bear dangling from her small hand. He turns and speaks to her.

We'll try to find something near Grandma, okay?

His daughter seems not to hear him, disregarding him, as if she doesn't care to listen. She turns away to look at something else, then takes her father's hand and pulls him to the window, points up at the night sky.

I think that star is Mommy, she says. It's the brightest.

We cut to a new scene now, later on that same night. The man is back on his computer, using Homesteader to create highly individualized preferences and parameters. His features soften. He smiles at something unseen.

New scene. In the warm vestibule of a new house, the girl runs into her grandmother's waiting arms. They are happy.

That night on the front porch, the father gazes up at the stars, his daughter standing on the lawn before him, her hands raised to the sky. The camera zooms in on the brightest star above.

Good night, Mommy, the girl whispers.

■ ■ ■

When I arrived I hesitated out front for a while. I stood shivering in the driveway, listening to the walls of dark pines breathing around me, watching for signs of life in the old house. I took in the dark gabled windows, the missing shutters, the rotten columns of the porch where I once lay sobbing and powerless as my father had his way with my sister. From where I stood I could look in and just make out the television in the living room, its weak and stuttering light flitting against the walls like a trapped bird. I looked up above me, but I could see no stars—the sky had clouded over, paling to a warm, unearthly mauve. Soon it would begin to snow.

In that moment, I decided that I should leave my daughter behind. I knew that she would be safe here, surrounded by the strange and broken women who loved her. They may have had their flaws, but they would do their best to take care of her. They would chip away at her inchoate anorexia, fattening her up with their relentless cooking and feeding. They would launder her uniforms every Sunday and see that she stayed at Hutch, where her clan of girlfriends would close ranks around her, precious survivor of unspeakable tragedy. Teachers would go easy on her; therapists would help her get through the worst of it. And perhaps one day she would write about what had happened to us, just as beautiful, brave Evie had intended to write her own sad story, before she gave up and jumped.

The truth was that I didn't deserve to see my daughter grow up and become a woman. Not after what I had done to us.

But who would take care of them if I were gone?

No, I thought. No. Whatever might happen to us now, my daughter belonged with me. There was no truth deeper than that.

I let myself in through the unlocked door, my daughter's coat slung over my shoulder, its soft sleeves dangling against my chest. In the front hall I turned away from the murk of my reflection in the credenza mirror, and with careful quiet steps I passed through the French doors and into the lightless living room where my sisters lay lumped together on the sofa like dogs, snoring in guttural wheezes. On the television screen two teenagers in parochial school uniforms—a black boy and a blond girl—were arguing fiercely, their attractive, tearstained faces drawn close, as if to kiss. My mother was nowhere to be seen.

If anyone else had been awake—if I'd had to speak with my sisters, or see my mother—I still may have given it all up, even then. But the only soul stirring was my girl, huddled in the rose-patterned wingback chair that dwarfed her little body, her dark eyes glazed with sleep as she gazed at the escalating fight on the television screen. I stood and watched her for a while. Even then—even then—I still could have stopped.

I whispered her name.

She looked up. Saw me. I beckoned to her with a nod, and she got up out of the chair and approached me, yawning widely. I put my arm around her, draped her coat around her shoulders—together we walked out of the house, and my daughter let out a little gasp.

It's *snowing*?

It was. We stood and watched the flakes drift through the streetlight, specks of static falling and falling through the night. It fell on our shoulders and hair, too soft and light to chill us. The silence was smothering and complete, thick against our ears. My daughter did a little twirl, as if she were playing the role of an enchanted girl, and I thought I saw Evie again, performing for us—for me—as she always had.

In the car I made sure my daughter buckled up. The billy club was still resting on the backseat, hidden in a lean-to of shadow. As I drove away I looked back at the house once, briefly. I was reasonably sure that I would never see it again.

My daughter fell asleep beside me almost immediately, and as I accelerated westward down the abandoned road I let my thoughts race on unencumbered. It was just after midnight—I could drive all through the night, and by early morning we would be far away. The Canadian border was only seven or so hours north, and there would be no traffic. Perhaps we could go to Montreal—I had never been there myself, but I'd heard good things. I hadn't packed anything for us, but it didn't matter. I had my wallet, so we would be able to buy whatever we needed for our new start. I imagined a tree-lined street in a new city, a bilingual girls' school, two pairs of scuffed snow boots waiting by our back door.

The worst was over now. It would be hard to move forward, but in the end we would be all right.

But I hadn't taken our passports. They'd been right there, stashed in the drawer of my bedside table, but I hadn't thought to take them with me when I left the house. I hadn't been thinking at all.

I looked over at my daughter, and then I knew that there were other problems, insurmountable ones that I hadn't let myself face until now.

What would I say when she asked for her mother? What could I possibly tell her? I knew in my heart that there was nothing I could say, that no possible explanation would make it all right. And now it was much too late to bring her back.

I had changed my mind about that, anyway. I no longer saw that house as a safe haven for my girl. She would wither away there. Her life would be nothing like what I had wanted for her. My family didn't have the means or wherewithal to take care of her. They couldn't even take care of themselves. Soon my ailing mother would die, and my sisters would fully capitulate to their trash mysticism and squalid sloth. Money would run out, and my daughter would have to go to the unfathomably lousy district school, if my sisters even bothered sending her to school at all. She would end up just like the twins, stunted and perverted, shut into that crumbling sanatorium for the rest of her life, which would really be no kind of life at all.

Or she would end up like Evie, reckless with a rage that she would never fully understand, stumbling headlong toward drugs and johns and her own self-immolation, forever mauled by her trauma.

That couldn't happen to my daughter. I couldn't bear to imagine her alone in a world so full of peril, without me there to protect her. She couldn't live like that, failed by everyone who loved her.

She couldn't live knowing that I was the one who had killed her mother.

I didn't think anymore. I was calm. Blank. I made the turn onto the parkway, drove us over the bridge. I knew there was water down there, churning below us, but it was too dark to see it.

At the beach I parked close to the rail, just behind the wall of the dunes. I put out my hand and stroked my daughter's shoulder, watched her stir. She let out a little moan as she peered into the impenetrable darkness outside her window. Then she turned to me, her eyes still squinted shut, still half asleep.

Where are we?

Come on, I whispered. Let's walk for a little while.

She was too tired to resist, to ask any more questions. She was my daughter, and she trusted me.

It would all be over soon.

We got out of the car, walked a little way out onto beach. I reached for my daughter's hand and we looked out over the white-capped waves, the snow dancing around us. She yawned, leaned her head against my chest. I held her close as she shivered.

Daddy?

What, love?

I'm tired.

Then lie down, I said.

Right here?

Sure, I said. Here, I'll lie down with you.

We lay down together, shoulder-to-shoulder. The sand still held a hint of warmth from the heat of the day, even as the snow fell soft on our faces. We gazed up, watched the blinding swirls come down on us from the dark, melting before they reached the ground.

Can we go home?

Soon, sweetheart, I promised. Soon. Isn't this beautiful?

She didn't answer—she had fallen asleep again. Very gently, I turned her onto her side and covered her as much as I could with my coat.

I walked to the car and retrieved the club from the backseat. Then I hurried back. The last image I have is of a dark mound on the sand. The back of a dark head.

Everyone uses the photograph of us on the beach.

It's sunset, and the flash has turned us into ghosts against a red sky.

My wife is farthest away, wrenching herself toward the water, her slender arm raised in what looks like a wave of farewell.

My daughter is between us, her hands gripping ours, holding us together.

And I'm closest to you, my chin raised in confidence and my eyes bright with pride, trying hard to look like a good man.

That was me. That was my family.

Club in hand, I walked quickly across the windswept beach, but not so quickly that I could hear my breath, or the beating of my heart.

It was a mercy that I only had to hit her once.

Acknowledgements

Thanks to my agent, Caroline Michel, and the great team at the PFD agency. To my UK publisher, Rachel Cugnoni, and my editor at Vintage, Ana Fletcher, who just kept pushing. To Elisabeth Schmitz at Grove Press, USA – the women who were with this project from the start.

Thanks to everyone who chipped in with useful things like Boolean algebra (Paul Shearer), and to Joss Kelvin, a digital innovator who knows the value of telling new stories.

And to Laura Evans, who copy-edited the manuscript (often), looked after my garden, and looked after me.

penguin.co.uk/vintage